The WAY TOYS WORK

The WAY TOYS WORK

The Science Behind the Magic 8 Ball,
Etch A Sketch, Boomerang, and More

Ed Sobey and Woody Sobey

CHICAGO
REVIEW
PRESS

Library of Congress Cataloging-in-Publication Data
Sobey, Edwin J. C., 1948–
    The way toys work : the science behind the magic 8 ball, etch a sketch,
boomerang, and more / Ed Sobey and Woody Sobey.
        p. cm.
    Includes bibliographical references.
    ISBN 978-1-55652-745-6
    1. Mechanical toys—History. 2. Toy making. I. Sobey, Woody. II. Title.

TS2301.T7S63 2008
688.7′28—dc22

                                                            2008001303

Cover and interior design: Scott Rattray
Cover photos: iStock.com

First edition
Published by Chicago Review Press, Incorporated
814 North Franklin Street
Chicago, Illinois 60610
ISBN 978-1-55652-745-6
Printed in the United States of America
5 4 3 2 1

To everyone who plays with toys and wonders how they work

CONTENTS

ACKNOWLEDGMENTS

Bob Guildig and the staff of Eastside Trains, Inc., in Kirkland, Washington, introduced us to the new era of high-tech electric trains. We're still running the American Flyer train Ed's father bought for him when he was born. Bob demonstrated integrated circuit remote-controlled trains that smoke when you press a button. Thanks, Bob.

Carl Kadie, who has helped with several of our books, loaned us his collection of Hula Hoops. Dominique Avery provided her expertise to demonstrate how to keep the hoops going. Joshua Wickerham took the photo of the Furby.

Matt Pierce and Don Uchiyama coached us on electronics issues. Bill Hones at Fascinations gave us the background on the AstroBlaster. Andrew Kamondy at Spin Master sent us the information on the invention of Air Hogs and a Liberator—wow, what fun that is!

Scott Eberle, Vice President for Interpretation at the Strong National Museum of Play, home of the National Toy Hall of Fame (which was founded by Ed Sobey) wrote several pieces that appear in the book. Rollie Adams, Director of the Strong Museum and National Toy Hall of Fame, wrote a piece as well. Rollie also supplied the photos of die-cast toys. Thank you, Rollie and Scott.

Thank you all.

INTRODUCTION

Wow, That's Neat!

Some toys command interest and inspire wonder. They do the unexpected or the seemingly impossible. They make us think about how they work and how they relate to the scientific concepts we've learned. They're just neat.

Fling an Aerobie and watch it fly, and fly and fly some more. Experience tells us we shouldn't be able to fling something so far with so little effort. Launch a boomerang with a hefty throw and it returns. That's crazy; things we throw don't come back. Draw a picture on an Etch A Sketch and erase it with the flip of a wrist so it's ready for the next creative idea. Pull a toy car back eight inches and watch it zoom 20 feet across the kitchen floor. What convoluted laws of physics allow for such seemingly magical experiences?

> I think we should teach them wonders. . . . The purpose of knowledge is to appreciate wonders even more.
>
> —Richard Feynman, physicist (1918–1988)

The magic of toys is not shrouded in secrets, but is there for us to see. Take screwdriver in hand and open toys up to enjoy the magic in a new way. While the toys themselves inspire awe, seeing how they're made lets us admire the engineers' creativity and problem-solving abilities. As awesome as the toys are to play with, they are even more awesome to understand.

Toys of technological wonder provide a common ground for learning and fun. This is the type of learning that comes from experiments, discovery, and experience—learning that allows us to apply what we already know to help us understand what we see but can't fathom. This is authentic learning that can be applied to the world around us, and that can inspire more discovery and learning. *The Way Toys Work* is your launching pad for exploring awesome toys and discovering how they work.

Guidelines for Reverse Engineering

Reverse engineering is the process of taking stuff apart to see how it works. Governments and corporations do it all the time to reveal the technologies their competitors have invented. In the same way, you can reverse engineer toys to figure out how they do the awesome things they do. But before you start, let us give you some guidelines to follow.

You'll need some tools. A good Phillips screwdriver is essential. You might want to have several of varying sizes. Amazingly, the Phillips screwdriver on a Swiss Army knife fits more screws than any other screwdriver we've found. A set of jeweler's screwdrivers will let you open up many small devices. You'll want a few flat head screwdrivers, too. These few tools will open up many of the toys. However, other toys will require a hacksaw and needle-nose pliers. For toys encased in plastic, we use a rotary cutting machine.

You'll need stuff to take apart. Thrift stores and garage sales provide a steady source of stuff—and usually the price is right. Let your neighbors or friends know you're looking for formerly working toys, and they'll turn up some gems. In this throwaway culture we live in, there is always a supply of yesterday's technology and toys.

You'll need to take safety precautions. Wear protective glasses or goggles whenever you're taking stuff apart. Parts can fly out when you least expect them to, so always protect your eyes. Resist the temptation to wrestle with a toy that refuses to open up. Rather than getting physical with the inanimate object, outfox it. Look to see what is holding the pieces together; find that last screw you overlooked. Sometimes manufacturers place screws under labels or otherwise out of sight.

If you decide to force a toy apart, be mindful of what is downstream of the screwdriver. Anything that's in the "line of fire," or downstream, of your screwdriver—your hand, a nice table, your friend's face—will become its unintended target when you slip. Aim your force toward a deserted volume of space.

If you're taking apart a toy that plugs into an electrical outlet, cut off its electric plug before you begin your work. Then bend the plug's prongs outward (so that no one can plug it in) and throw it away. We didn't always do this—until the day a fourth grader in one of our programs inserted a discarded plug into a wall outlet. The resulting explosion left him shaken but unharmed, and the electrical outlet destroyed. Now we always bend the prongs outward so they can't be inserted into an outlet.

You'll need to sharpen your sense of awe. Awe sharpening is harder than awl sharpening, because you'll never get it sharp enough. In our haste to get the important stuff done, we too often overlook the stuff that makes life fun. We tell students that the number one rule of science is that when you find something interesting, stop and focus on it. So our plea, as much to ourselves as to you, is to never let the awe get dull. Be on the lookout for those things that might prompt you to think, "Wow, that's neat. I wonder how that works." Then go find out.

Patents

Isn't the Internet great? You can find out so much information while sitting in your pajamas at your computer. For instance, if you can find a patent number on a toy, you can see a copy of the patent online. It used to be an expensive task to find patents. Now there are several services that can retrieve them. We prefer www.google.com/patents. A patent describes, sometimes in painful detail, how the toy or other gizmo works. It contains drawings showing the parts, and it lists who filed the patent and when it was filed and granted. If you have a favorite toy, you might find it fun and informative to look for the patent number and go on a search.

Build Your Own

We are big believers in the idea of learning by making stuff. We also advocate innovating new products by creating prototypes. Making new stuff or improving existing stuff isn't a case of perfecting the design on paper, then going to the shop to build the final product. Before you build something you have to have *some* idea of how to make it work, but the development process will continue as you start to assemble the components. You'll learn so much that you didn't know you needed to know—you'll learn stuff that you would never have thought of just by looking at a two-dimensional design. Only when you start to put the pieces together do you find out if they fit and if your innovation will work.

Building inherently invites collaboration. People get interested, even excited, when they see what you're working on. You won't inspire that much excitement by just talking about a design or showing sketches. Prototypes beg people to get involved and to contribute. And each person's contribution can improve the final product.

So try building your own versions of these toys. Building and testing them is as much fun as playing with the ones you buy at the store, and you'll learn so much more.

AEROBIE

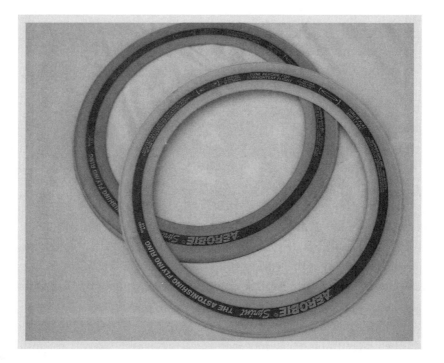

History of the Aerobie

Lots of toys are invented by someone stumbling onto a design or idea. Not so with the Aerobie. Inventor Alan Adler worked for years to perfect the design and find the right materials to create this flying ring, which can be thrown several times farther than a flying disc such as a Frisbee. Adler, an engineering professor at Stanford University, used his knowledge of aerodynamics to design the Aerobie. Before he invented the Aerobie he invented the Skyro, which set a Guinness World Record in 1980 when it was thrown 857 feet. The Aerobie beat that record when Scott Zimmerman chucked one 1,257 feet in 1986. Scott's throw set the world record for the longest throw of an inert, heavier-than-air object. Although the record was bested a few years later, it remains the world record for an "object without any velocity-aiding feature."

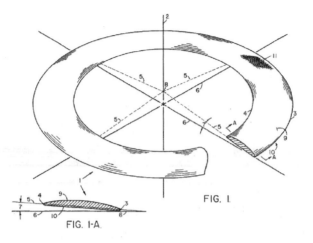

FIG. I.

FIG. I-A.

Patent no. 4,456,265

How Aerobies Work

To understand how a flying disc or ring flies, read about the Frisbee (page 52). Aerobies are different from Frisbees in several ways. Whereas a Frisbee has a blunt edge to create turbulence and reduce lift at the leading edge, the Aerobie has a unique lip, or spoiler, on its outer edge to create stability. Its slender profile presents much less drag surface than a flying disc. Less drag allows it to travel farther.

The Aerobie also differs from the Frisbee in that, as a ring instead of a disc, it has an opening in the center. In flight, air moves through the opening so that the inside edge is also a leading edge, which generates more lift.

You can make your own version of the Aerobie and determine how much of a spoiler is needed to get a good flight.

Inside the Aerobie

By looking at a cross section of the Aerobie, you can see features that may have otherwise eluded you. Cut through one side of the Aerobie, as shown here. A coping saw or another saw with a slender blade works well.

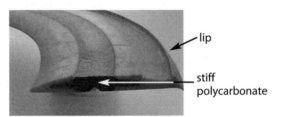

lip

stiff
polycarbonate

Check out the shape. It looks like a wing, which it is. The odd part of the design is the lip, or spoiler, on the outside edge. That spoiler was the breakthrough design feature.

Another problem Professor Adler had to overcome was making the outer edge soft enough that it wouldn't hurt someone who might be hit by it, yet keeping the toy rigid enough to fling. That's why you'll find a dark polycarbonate backbone in the middle of the softer and lighter outer material.

If you want to put your Aerobie back together, glue the cut edges back together with plastic glue. Reinforce the weld by gluing a craft stick to the underside, across the glued edge. It looks a bit odd, but it will fly fine.

Build Your Own

Cut out the centers of several thick paper dinner plates. Fling your "rings" to see how they fly. A few brands of paper plates are made of heavier material; they make for rings that fly well without modifications. Most paper plate rings, however, require some help.

Mount two plates upside down on top of a third, right-side-up plate to form a flying saucer–like toy, and tape them together. Check how this flies. To improve the flying characteristics, try affixing four metal washers or pennies at even intervals along the outer edge. Now try out your creation. Next, try bending up the edges of the plates to make a lip like that of the Aerobie. Adjust the position of the lip until your ring flies level. Then try adding more evenly spaced weights to your ring. The goal is to create a ring flyer that travels far. It should fly level, without either side rising.

You can also create a ring flyer by cutting out the center of an empty pie pan. Adjust its flight by bending the outer edge up or down.

Resources

Teachers wanting to incorporate Aerobies and Frisbees into classroom learning opportunities should look at Ed's book *Loco-Motion: Physics Models for the Classroom* (Zephyr Press, 2005).

AIR HOG

History of the Air Hog

The idea of a toy plane that's driven by air pressure and a piston came from inventors in England. The inventors were unable to sell their idea to big toy companies—they hadn't yet made a flying prototype—but they found interest in a young Canadian company, Spin Master. Following the success of two previous toys, Spin Master invested a half-million dollars to develop the idea into a product. Beginning in 1998, sales—like the planes themselves—took off.

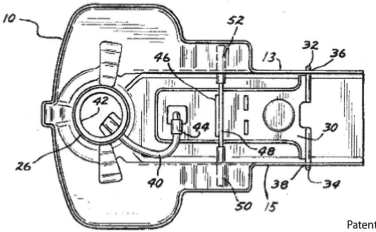

Patent no. 3,789,540

How Air Hogs Work

Like all Air Hogs toys, an Air Hogs car has a pneumatic motor. Air pressure—you provide the energy by pumping air into a pressurized reservoir—pushes on a piston. The piston is attached to a crankshaft that converts the linear motion of the piston into the rotary motion of the wheels. Add some gearing and a flywheel to keep the spin going through angular momentum, and the car is ready to go. To control air flow to the piston's cylinder, there are intake and exhaust valves. Each has a tiny black ball that you can see (and easily lose if you take the car apart!).

The car comes with a pump to fill the reservoir. Air enters the reservoir through a ball valve, supported by a spring, that lets air into the reservoir, but not out. When the reservoir is filled and you're ready to play, you spin the left rear wheel (or the propeller, if you have a plane or helicopter). As it rotates, it turns the crankshaft (attached through the gears). The crankshaft pushes up on the piston, which is connected to a spring. The spring lifts the bottom of the second valve to let air into the chamber, driving the piston down.

The crankshaft is attached to a metal gear with eight teeth. It drives a plastic gear with 22 teeth, and that gear drives a third gear with 52 teeth that is attached to the wheel. When a small gear drives a larger gear, which in turn drives an even larger gear, these gears in series slow down the rotation speed (from small to large) but increase the torque, or turning force, that powers the wheels.

Once we took apart the car we were able to estimate that for every turn of the big wheel, the piston completed six or seven cycles. This estimation was validated when we calculated the gear ratio, which is $52/8$, or 6.5. (The biggest gear has 52 teeth, and the smallest has eight. Therefore, for every revolution of the wheel, the piston moves up and down 6.5 times.)

Inside an Air Hog

The Air Hogs car is a great toy to take apart. Turn it over to see the transmission. Pushing the lever to the "Power Sprint" side engages both rear wheels to the power train, making the car go straight ahead. Sliding the lever to the "Spin Out" side disengages the right rear wheel so that the car, powered only by the left rear wheel, turns in circles.

reservoir (filled with colored water so you can see it)

engine

Using a Phillips screwdriver, you can loosen and remove the four screws that secure the body to the chassis. Next, disconnect the air reservoir. WARNING: Although you may be tempted to just twist off the air reservoir (clear plastic ball), *don't*. The reservoir has weak walls that you might damage by squeezing too hard. Instead, use pliers to twist the black collar in order to remove the reservoir.

Turn the car over to remove the four screws that hold the rear wheels and motor in place. With the air reservoir disconnected, you can pull the rear wheel assembly down from the chassis.

Look, from the side, at the valve that lets air in. You'll see a tiny ball supported by a spring. Air from the pump depresses the ball slightly, which allows air to flow into the reservoir. During the pump's recovery stroke, the spring pushes the ball up in place to prevent air from leaking back out. Once the spring has pushed the ball up, air pressure from the reservoir (in addition to the spring) holds it in place.

Slowly spin the left rear wheel. You'll see a white plastic piece—a piston—rise and fall six or seven times with each revolution of the wheel. Air can't escape the reservoir until the piston rises. To get the car to move, you turn the wheel, which moves the piston, which releases air. As the wheel turns, it moves the white piston up to push on another tiny ball that's attached to the large spring. When this ball moves, air escapes from the reservoir and pushes the piston down. When the piston moves down, the spring pushes the ball

valves

reservoir

back in place to block the flow of air. So once you start the wheel spinning, the reservoir delivers a blast of air to the piston six to seven times per wheel revolution. You can hear the "puff, puff" of air escaping when the car operates.

Lightly screw the reservoir back onto the valve assembly; this will hold the two parts of the assembly together. Two tiny screws hold the valve assembly onto the motor. Remove these with a jeweler's screwdriver. Now ask yourself: "Do I feel lucky?" If your answer is no, don't take apart the valve assembly. If your answer is yes, spread a white towel on your workbench to catch the two tiny balls that will soon come bounding out.

Three screws hold the two halves of the valve assembly together. Remove them, then unscrew the reservoir. That will allow the two halves to come apart. Watch one of the balls bounce away as you separate the halves. It'd be tough to find a replacement for one of these, so don't lose it.

Notice the O-ring in the upper half of the valve assembly. O-rings are used in high-pressure devices such as scuba tank valves and space shuttle tanks. Higher pressures squeeze the O-ring, causing it to make a better seal.

Now, before someone bumps the table and those tiny balls are lost forever, put the valve assembly back together.

To access the motor, unscrew the four screws holding the top piece. Lift out the crankshaft (the white plastic piece attached, off center, to the wheel), the metal gear, and the flywheel assembly.

When assembled, the crankshaft fits under the piston. When it pushes up on the piston, the piston lifts the lower ball, which admits air that drives the piston down and shuts off the supply of air. To start the engine you have to spin the wheels to initiate the cycle.

flywheel

cam

piston

The flywheel (the metal disc at the end of the shaft) keeps the shaft spinning. Once it has started to spin, its momentum will keep it rotating, allowing the crankshaft to release more air. (See Friction Car, page 49, for a discussion of another application of a flywheel.)

Finally, you can remove the several screws holding the two halves of the transmission housing together. While the housing is open, count the number of teeth in each of the gears. Now, before you misplace one of the parts, get the transmission back together.

The pump looks well constructed, but is otherwise unremarkable. We recommend leaving it intact.

ASTROBLASTER

History of the AstroBlaster

A conversation among physicists at a cocktail party was the genesis for the AstroBlaster, a stack of bouncing balls capable of rebounding to five times its drop height. Bill Hones, his father Edward, and Stirling Colgate share the patent (patent number 5,256,071).

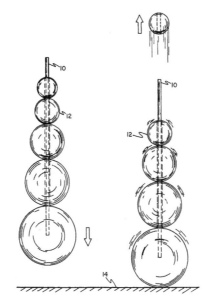

Patent no. 5,256,071

How AstroBlasters Work

According to the patent, the AstroBlaster demonstrates "an unobvious consequence of fundamental laws of physics—the acceleration of an object to high speed by multiple collisions among a series of heavier objects moving at slower speed."

We ran some experiments to understand what's going on. We pulled the shaft out of the bottom ball so we could bounce test it. We dropped each ball to determine its rebound. We estimate that each of the three bottom balls rebounds to about 75 percent of its drop height. Then we tried the top ball. *Kerplot.* It went almost nowhere, exhibiting about a 20 percent rebound. It seemed that the top ball was made out of a different material than the other three—but we would later find out that this wasn't the case.

When you drop the AstroBlaster, its collision with the ground causes some loss of energy, but not much. Physicists say this is an elastic collision—the total kinetic (moving) energy is the same before and after the collision. Of course, it's not exactly the same. Based on our rebound testing of a single ball, we estimate that it's about 75 percent; the other 25 percent of the kinetic energy is transformed, most into heat energy and a bit into sound.

When the stack of balls hits, the bottom (most massive) ball transfers its considerable momentum to the ball just above it. The second ball now has its own momentum plus the momentum of the bottom ball. When it hits the third ball, it transfers its enhanced momentum to it. The small ball on top is on the receiving end of this substantial transfer of momentum. Since momentum equals mass times velocity, and the top ball is much less massive than the ones below, the transferred momentum manifests as a much higher upward velocity—several times the ball's downward velocity, in fact.

But there's more. The maximum height the small ball can reach is determined by its kinetic energy, and that depends not on its velocity but on the *square* of its velocity. So the severalfold increase in velocity yields a significant kinetic energy change and a much greater rebound height.

But when the balls are bounced separately, why does the top one rebound so much less than the others? We tried to figure this out, and ended up calling one of the inventors. It turns out that all of the balls are made of the same material—polybutadiene, also used in SuperBalls—but the hole in the small ball makes its collision with the floor much less elastic. More energy is lost in the collision, so it rebounds less.

Build Your Own

The AstroBlaster features a great design—a post keeps the balls aligned and allows you to drop them so that they fly straight up. In making your own version of the AstroBlaster you'll lose this property, but you'll still end up with a neat toy.

You need a basketball, a tennis ball, and a lighter ball such as a racquetball. Find a clean tin can that's just a bit larger than the tennis ball and remove both ends. Duct tape one end of the can onto the basketball, and you're done.

To operate it, go outside—this step is important for maintaining domestic tranquility—and put the tennis ball into the open end of the can. Hold the apparatus so that the can points straight up. (This is the time to mention what a great investment safety goggles are.) Release the basketball and watch the tennis ball fly when the basketball hits the ground. Repeat the experiment, this time placing the racquetball on top of the tennis ball. Whoa! Just how tall a stack of balls can you create?

Science Experiments

To convince yourself that energy and momentum are transferred from the basketball to the other balls, measure the rebound height of the basketball with and without the other balls. You'll find that the basketball has a lower rebound height when it transfers momentum to the other balls.

Resources

Fascinations (800-544-0810) sells the AstroBlaster, as do several science catalogs and Web sites.

For a more complete description of the physics behind the AstroBlaster, see *Turning the World Inside Out and 174 Other Simple Physics Demonstrations* by Robert Ehrlich (Princeton University Press, 1990).

BALSA WOOD PLANE

History of the Balsa Wood Plane

The largest manufacturer of balsa wood model planes got its start in a barn in 1926. Paul Guillow founded his toy company just as aviation was becoming popular. The year after Guillow launched his company in Wakefield, Massachusetts, Charles "Lucky Lindy" Lindbergh made the first solo transatlantic flight. Lindbergh's success spawned interest in model planes, and Guillow's business took off. His first models sold for 10 cents. The company continues to make balsa planes, but today it also makes foam and plastic models, kits, and promotional materials.

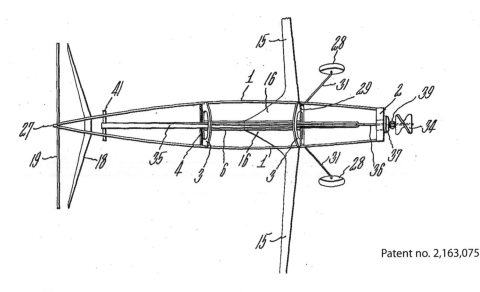

Patent no. 2,163,075

How Balsa Wood Planes Work

Balsa gliders are tossed by hand into the air or launched with a rubber band. The propeller-equipped model discussed here has a rubber band that is wound up. Of course, it works only if you wind it up in the right direction. Look at the propeller to figure out which direction of spin will catch air and push it toward the rear to move the plane forward. The propeller must be wound in the *opposite* direction. Or take the trial-and-error approach: wind the propeller up and let it spin to find what direction it moves the air.

The propeller pulls the plane forward by pushing air backward. This demonstrates Newton's third law: for every action there is an equal and opposite reaction.

Amazingly, it's been more than a century since Orville and Wilbur took their first flight, and scientists still can't agree on how planes achieve lift. If you're a fan of mathematician Daniel Bernoulli, you won't like our explanation. Bernoulli fans tell you that when two air molecules separate upon encountering the leading edge of a wing, they have to rejoin at the trailing edge. Since the molecule traveling over the top of the wing has a longer path (due to the curved shaped of the wing), it must move at a higher speed. And in a closed pipe (clearly not the environment a plane is flying in), those higher speeds would be associated with lower pressures. By this questionable reasoning, the wing is pulled upward by the decreased pressure above it.

There's a much cleaner way to describe how lift is achieved in both normal and inverted flight. As the plane moves forward, the wings encounter air molecules, which tend to follow the curvature of the wings—an example of the Coanda effect. The wings deflect the air downward, and the equal and opposite reaction pushes the wing and plane up. There's Newton again.

With no acceleration (constant speed and steady altitude), a plane's weight is balanced by its lift. Its aerodynamic drag is balanced by the thrust created by the propeller. Of course, flying a balsa wood plane gives you darn few seconds of constant speed and steady altitude.

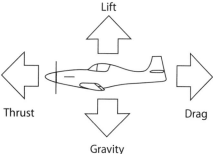

The plane's tail provides directional stability. The vertical stabilizer keeps the plane from "yawing," or turning from side to side. The horizontal stabilizer keeps the nose from "pitching," or moving up and down.

Build Your Own

We recommend that you purchase a balsa wood airplane kit, but you can also make your own model from the throwaway styrene container that you purchase with a pound of hamburger or steak. First, wash and dry the white plastic. Next, cut off the sloping sides to yield a rectangle of airplane modeling material.

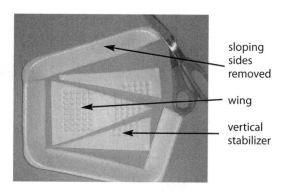

Cut the largest isosceles triangle possible out of the rectangle, as shown. In other words, find the center point of one short side and cut from that point to each of the distant corners. This will give you one triangle (about to become the wing) that has equal length sides and two right triangles (one of which becomes the vertical stabilizer).

Cut a slot in the center of the back of the wing. The slot should be just wide enough for the soon-to-be created vertical stabilizer to fit through it, but hold it tightly. Cut it about one-third of the length of the wing (toward the nose or tip of the triangle).

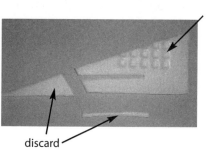

Cut an inch off the small end of one of the two right triangles, then cut a slot in it, parallel to the long leg of the triangle about as long as the slot in the wing.

Slide the vertical stabilizer into the wing, slot sliding into slot. Put a paper clip in the nose of the plane (the paper clip adds weight to the nose and prevents it from rising) and you're ready to take off. If, after you try flying the plane, you find that the nose still rises, just add one or more additional paper clips.

You may want to outfit your plane with ailerons, the movable wing surfaces (invented by Alexander Graham Bell) that allow a plane to bank. Cut flaps into the back of the wing and try raising both, lowering both, or raising one and lowering the other. Cut a rudder into the back of the vertical stabilizer and experiment with that as well. The big challenge is to get your model to do a banked turn.

Next, try adding rubber band power. You'll probably need a new model, so eat another hamburger or two and prepare the styrene. Loop a rubber band onto the end of a 10-inch-long, ¼-inch-diameter dowel. Tape and hot glue a bent paper clip onto the bottom of the wing about an inch from the nose. (WARNING: hot glue melts the styrene, so use it sparingly.) Loop the rubber band onto the paper clip. Pull the dowel forward while holding the plane, then release the plane. Now you're flying.

Science Experiments

One of the greatest things about balsa wood fliers is that you can ask, "What happens if the wings are moved forward?" and seconds later you can run an experiment. Ask, "Why does the plane need a vertical stabilizer?" then pluck it off and launch the plane. Talk about getting to know the elements of flight!

Resources

Understanding Flight by David F. Anderson and Scott Eberhardt (McGraw-Hill Professional, 2000) is a great book for helping to clarify the physics of flying.

For classroom projects using this model, see Ed's *Inventing Toys* (Zephyr Press, 2002).

BICYCLE

History of the Bicycle

Bikes have a long and interesting history, from early designs that didn't include pedals up to today's lightweight carbon fiber frames designed to go as fast as possible. In Paris in 1818, Baron Karl von Drais publicly demonstrated his new *Laufmaschine*, or running machine. It was the first two-wheeled, human-powered, steerable form of transportation. The invention became a fad across Western Europe and North America for a few years, with many people taking an interest in engineering and mechanics and building their own versions (which were commonly called velocipedes, meaning "fast feet"). Most riders preferred to ride their velocipedes on sidewalks instead of on the streets, which led to many accidents with pedestrians. Because of this, several cities banned their use.

Between the 1820s and 1850s, the two-wheeled models went out of fashion and were replaced by tricycles and quadricycles. Inventors started experimenting with different methods of powering these vehicles, including adding pedals and hand cranks. Eventually, however—thanks in large part to the cumbersome weight of the extra wheels and heavier frames—people started going back to the two-wheeled designs.

"When H. G. Wells, the dark prophet of science fiction, imagined the shape of things to come, he saw many frightful scenarios: invading martians, human evolution gone awry, world anarchy, and nuclear chain reaction. But he imagined shining outcomes, too. 'When I see an adult on a bicycle,' he wrote, 'I do not despair for the future of the human race.' When we're all grown up many of us hang onto our bicycles, optimistically. We admire the efficiency of the bicycle, approve of the way it saves energy, and love using it to make better time than cars in heavy traffic. But these grownup virtues are side issues. Bikes are about play: balancing on two wheels, rolling downhill, passing huffing joggers, splashing through a puddle, and smiling. The bicycle is the toy that lasts."

—Scott Eberle, Vice President for Interpretation,
Strong National Museum of Play,
home of the National Toy Hall of Fame

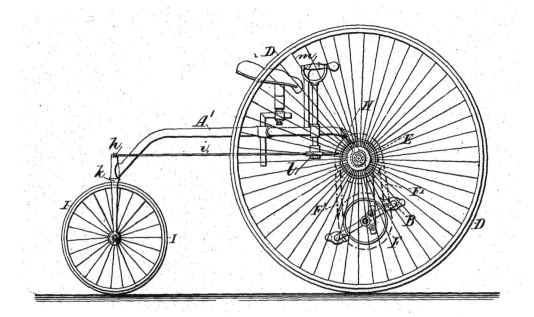

Patent no. 274,231

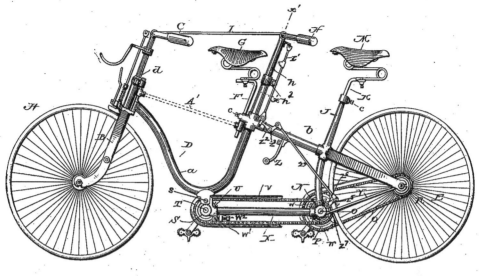

Patent no. 415,072

In the 1860s the first popular pedal-powered, two-wheeled vehicles hit the scene. One of the first types was nicknamed "the boneshaker" because, with its wooden frame and wheels and iron bands as tires, it made for a very uncomfortable ride across cobblestone streets—but it was still faster than walking. A few years later, James Starley, a British inventor, became the "father of the bicycle industry" when he invented the penny-farthing. When most of us think about historic bicycles, we picture the penny-farthing, which got its name from the relative sizes of its wheels. It had a large front wheel and a small back wheel, which resembled two British coins, a penny and a farthing, placed side by side. This was the first bicycle to have rubber tires, which made for a smoother ride. Also, by maximizing the size of the front wheel, riders could achieve greater speeds than were possible with previous models of gearless bikes. Unfortunately, the larger wheel meant that the rider had to sit up higher and was more likely to go flying over the handlebars if he or she hit anything.

After incurring many injuries on the penny-farthing, Starley's nephew, John Kemp Starley, invented a less dangerous model, the safety bike. It would remain the standard bike design until the 1980s, when mountain bikes became popular. The safety bike had a diamond-shaped frame with two wheels of the same size and a chain drive from the pedals to the back wheel. With the addition of John Boyd Dunlop's pneumatic tire, not only were these bikes safer, but they also offered a smoother ride.

Recently, new materials have allowed inventors and bike enthusiasts to create new designs with extremely lightweight frames, stronger pedals and cranks, and more effective braking systems. All of these have allowed people to bike (and unicycle, tricycle, and quadricycle) places they couldn't just a few years ago.

How Bicycles Work

It's quite easy to see how most bikes move. The simplest bikes don't have gears; the spinning is transferred from your feet to the pedals, then to the crank arms, and then to the sprocket, which drives the chain to spin the rear sprocket connected to the rear wheel and propels you forward. Nothing to it, really—just a direct-chain drive. When you add changing gears, though, things get a bit more complex.

The complexities of movement aside, however—what keeps you *balanced* above two small points of contact with the ground? For many years people thought that the spinning of a bike's wheels provided the vehicle with gyroscopic stability—in other words, that the spinning wheels created forces that keep the bike upright. But this theory has since been debunked. Several people have done experiments in which they've mounted additional wheels on a bike and spun them backward to cancel the gyroscopic effect of the two forward-spinning wheels, and they found that they could still balance themselves. (You can re-create this experiment for fun, but you'll have to be inventive about how to attach the additional wheels.)

The main way of staying balanced consists of angling the front wheel to keep your center of gravity directly between the points at which each wheel contacts the ground. One way to demonstrate this is to prevent the front wheel from angling right or left; if it can't angle, you won't be able to stay balanced on the bike, even if you try to ride in a straight line. The only time the center of gravity doesn't have to remain above those two points is when you're turning. A bike that is in the process of turning experiences not only the force of gravity but centrifugal force as well. To stay balanced while turning, you must take into account both forces by leaning into the turn. The sharper the turn and the faster you take it, the more you must lean.

Inside the Bicycle

Bicycles can be very complicated machines, but most of the parts are visible without us having to take anything apart. The main components of a bike include wheels, frame, drive train, brakes, steering, and seat. Some newer bikes also include suspension parts to make the ride smoother. Each of these components is made up of several smaller parts that we can examine with special tools.

Wheels are the primary bike parts. Start your investigation by removing the front wheel. Usually it is held on with a nut and bolt, which can be removed with a wrench. You may have a wheel with a "quick release," in which case it can be removed with just your hands. The quick release was invented in 1927 by Tullio Campagnolo, an Italian bicycle racer, to help him repair flats quickly and win more races.

Campagnolo's design utilizes a cam, a wheel of uneven radius that is used in many inventions; as a cam rotates, the piece it pushes against gets closer or farther from the center of the cam. The quick release uses a lever arm to rotate a cam and tighten itself down like a nut; the cam tightens much faster than a wrench can turn a nut, since it doesn't have to be spun as many times. The addition of a lever increases the force on the cam and allows you to tighten your quick release even more quickly.

Once the wheel is off, you can see the hub that holds the bearing, the device that allows the spinning bike wheel to connect to the non-spinning bike frame. Without the bearing, the whole bike would rotate around one of the wheels—which, among other things, would be an uncomfortable way to ride a bike.

If you let the air out of the tire tube, you can remove the tire and inner tube from the rim of the wheel. The tire isn't all that exciting, but the tube is pretty cool. Basically it is a rubber tube with a valve that allows air to flow into, but not out of, the tube unless you press open the valve. (The coolest thing about bike inner tubes is all the other things you can do with them—see Stomp Rocket, page 135.)

The frame is an important part of the bike; it connects all the other parts together. New frame models are designed each year to decrease the weight of the structure while still providing strength and rigidity. Most bikes have frames that are made from tubular metal, either steel or aluminum. These metal tubes are welded together into triangles: a main triangle and two rear triangles. Such triangular arrangements are known as truss structures, and they're used because they make for strong, lightweight frames that can withstand the forces of biking.

Attached to the frame is the drive train, which allows the rider to transfer energy from the pedals to the rear wheel. The most visible parts of the drive train are the chain and the sprockets. One of the most complex devices on a bike is the derailleur, the mechanism

that allows the rider to change gears while riding. By shifting the position of the derailleur inward or outward, you force the chain to align with different sprockets.

Finally there are the brakes. There are three main types of bicycle braking systems used today. On many kid's bikes coaster, or back-pedal, brakes are used. The coaster brake has a clutch inside that pushes the wheel forward or allows it to coast until the rear sprocket is turned backward. When

The derailleur moves in toward the bike or away from the bike as you shift the gear lever. It forces the chain to move so it can engage each of the two or three sprockets.

Chain transmits energy to the rear wheel through the rear derailleur.

Front sprocket

that happens, the clutch inside is pushed in the opposite direction and forced between two brake pads to slow the wheel.

The other two types of braking systems are similar to each other, and it's easier to see how they work. Rim brakes have two pads that pinch the rim of the wheel, while disc brakes pinch a metal disc attached to the hub. Both are controlled by cables attached to the handlebar, and both use the friction against the brake pads to turn kinetic energy into heat and slow you down.

Build Your Own

Building your own bike from scratch would be a fairly complicated venture. First you'd have to weld together a frame. Then you'd have to cut the gears and construct the wheels. It would be a big challenge to make a perfectly round wheel by hand. Another approach is to go to a bike shop, buy all the parts you need to build a custom bike, and assemble it yourself.

BIG MOUTH BILLY BASS

History of Big Mouth Billy Bass

A Bass Pro Shop inspired Joe Pellettieri, vice president of product development for Gemmy Industries, to make a singing fish in 1998. After a few prototypes—the first was called "hideous" by the company's marketing VP—Gemmy Industries came up with Billy Bass. Glenn Robinson received design patent D440,525 in 2002.

Like most fads, it seems to have passed quickly. Billy must have gone back to the lake, because he doesn't swim or sing in stores these days. To get your own Billy Bass, check out eBay.

How the Big Mouth Billy Bass Works

Billy Bass has three DC motors that move its jaw, head, and tail. Springs return the fish to its stopped position. To turn Billy on, you either push a button or pass in front of its light sensor, which is mounted above and below the nameplate.

The circuit board coordinates the motion and the sound (singing) with a technology called Syncro-Motion. Four C cell batteries power all this fun.

Inside Billy Bass

We filleted Billy Bass to see what was inside. But if you want to be able to reassemble him so that he's practically good as new, you'll need to start from the back instead. Remove the several screws to lift off the back plate. The battery housing is part of the plate, and the speaker is attached to the plate. Unscrew the six screws that hold in the motor. This motor turns Billy's head away from the mounting board. You can slip the cloth out from underneath the screws so you won't have to cut it on the other side. We recommend putting back the screws to hold the motor while you work on the other side.

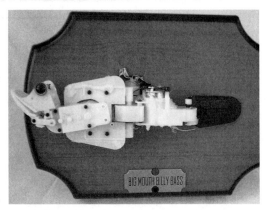

While poking around inside, remove the two screws that hold the switch below the motor. (You can also remove the on/off switch on the back plate if you wish.) Pull out the switch to see the tiny black conductor that you press to make contact across the switch's shiny terminals. Using pliers, you can wrestle out the "electronic eye," a photocell that senses changing light levels so that Billy knows to start singing when someone approaches. The eye is located on the bottom of the nameplate. The circuit board comes out by removing the two small screws.

Now for the filleting. Using a pocketknife, cut along the belly until you can pull the rubber skin over the tail. Then work under and around the head to cut the skin away. To get the skin over the eyes, pull. The eyes are secured with a compression fitting. To remove them, cut the rods holding them from the inside. (We couldn't stand the thought of Billy without eyes, so we left them in.)

Under the skin, small blocks of Styrofoam are glued to the mechanism to fill out the rubber body. Peel these off to reveal the plastic guts. Wow! If you like to take stuff apart, Billy Bass is great.

Remove the tail fin, which is secured by two screws. (Time out to mention that if you lay the screws down in a row as you take them out, you'll be able to figure out what size screw goes in each hole.) The lower support for the tail comes off with the tail itself, allowing us access to the gears that drive the tail.

The motor that swings the tail is mounted on top; it drives the gears, which are accessed from below. The axles supporting the gears are held between the upper and lower parts of the housing, so as you open the lower housing, be aware that axles and gears may spill out. (If you are confused about how to replace the gears, know that they are used to reduce the speed of the DC motor, so small gears drive big gears.) Rather than reverse the current flow to return the tail to its home position, the manufacturer has used a spring.

Notice that the motors each have diodes soldered across the terminals. The diodes protect electronic components from electricity generated by the motor. While power is driving the motor, the diodes don't conduct electricity. (Diodes conduct electricity in one direction.) When power is shutting down or starting up, diodes send any electrical "noise" from the motor back to the motor rather than to the electronics.

Before exploring further, we suggest that you reassemble the tail gears and close the tail section. To take the entire tail assembly off, remove four screws, two on top and two underneath.

To remove the head, first take out the screw that holds the return spring. It's located on the inside cheek. Now you can take out four screws and slide the head off the body.

To get access to the motor and gears that move the jaw, first remove the two screws that hold the top of the head. You can see that the jaw supports are blocking access to the gears. Remove only the top support. The bottom, or inside, support also holds the return spring—a minor hassle to replace, but a hassle you don't need.

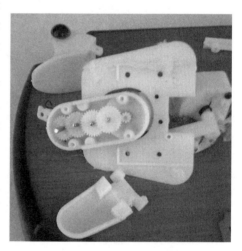

If you want to remove everything from the plate, return to the inside and take out the six screws that hold the motor and its mount. Now you can slide the motor through the opening.

This is a test. Can you get it back together, with no leftover parts and in good working order? Of course the skin may sag just a little, but that's OK. Billy won't mind.

Make Your Own

Making a Billy-like device from scratch would be quite a challenge. However, several Internet sites show you how to hack into Billy's system and retrain him to say what you want him to. One such site is at www.mit.edu/~vona/bass/bass.html.

BOOMERANG

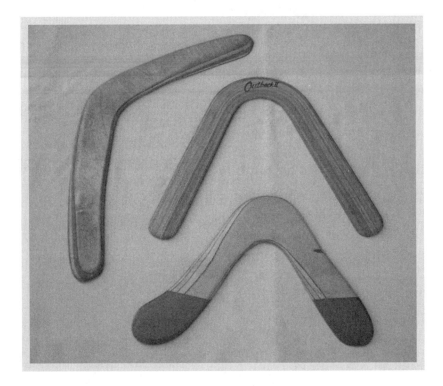

History of the Boomerang

Native peoples in the Middle East and Australia have long had boomerangs, and they probably used them to develop throwing skills. Boomerangs were found in King Tutankhamen's tomb in Egypt, and even older ones have been found in Australia. Although many cultures used throwing sticks as weapons or for hunting, the boomerang filled neither of these roles. It's our guess that young hunters developed their throwing strength with boomerangs (it's much easier to practice your throwing when you don't have to walk across the field to retrieve your stick) and that later in life they used throwing sticks (that did not return) in battle and hunting.

How Boomerangs Work

The ability to spin and an asymmetric shape are essential for a boomerang. The thrower holds the boomerang vertically by one end and spins it as he or she throws it. The wing-shaped surface of each boomerang arm creates lift. Lift on the upper arm is greater since it spins in the same direction the boomerang is traveling. Lift on the lower arm is less since it's moving the opposite way from the boomerang's direction of travel. In other words, air speed measured on the upper arm is the sum of the boomerang's forward motion plus the forward spin of the upper arm, but on the lower arm, the measured air speed is the boomerang's forward speed minus the speed of rotation. The greater the air speed, the greater the lift.

Because the boomerang is flying vertically, the increased lift on the upper arm causes the boomerang to lay over to its left side (assuming it's thrown by a right-handed person). In addition, the spinning boomerang's lift-generated torque causes the boomerang to turn to the left; this is the gyroscopic effect. This is the same effect you experience while riding a bicycle: if you lean to the left you put torque on the front wheel, but instead of falling to the left, the wheel itself turns to the left.

As the boomerang turns to the left in flight, it lays over so that it flies in a nearly horizontal position, and its lift causes it to rise. From the top of its trajectory, now about halfway around the circle, it falls back to the thrower. With some luck and skill the thrower can deftly catch (with one hand on top and one hand underneath) the still spinning boomerang.

Inside the Boomerang

Boomerangs are made of solid wood or plastic, so there isn't much to see on the inside. What is interesting, however, is the boomerang's shape.

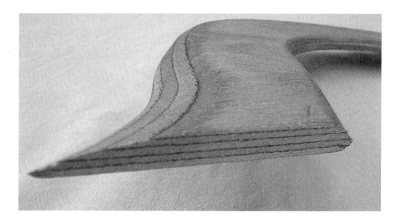

Almost every toy that's meant to be thrown is symmetrical. You can't tell one edge of a Frisbee, Aerobie, or baseball, from another. But a boomerang is asymmetric; you can see a left side and a right side. It's the asymmetry that allows it to return.

Check out the cross section of a boomerang arm. It looks like the wing of a plane. The steeply climbing side is the leading edge, and the other side is the trailing edge. This shape generates lift. Try flinging a piece of wood that doesn't have a wing-like profile, and you'll see not only that it won't return to you, but also that it doesn't travel very far.

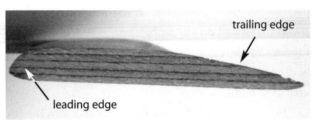

The other thing to notice is that, unlike an airplane wing, the boomerang has leading edges on opposite sides. The outside edge on one side is a leading edge, and on the other side it's a trailing edge. This is because, unlike airplanes, boomerangs spin to fly.

Build Your Own

Balsa boomerangs are simple and quick to build. Here's a way to make one in just half an hour. Cut two 11-inch-long, 1-inch-wide arms (1 inch by 11 inches) out of ⅛-inch-thick balsa wood or plywood. Sand the wings to create a 45-degree rise on the leading edges (on opposite ends of each arm), then sand a gradual descent to the trailing edges. The two arms should be identical.

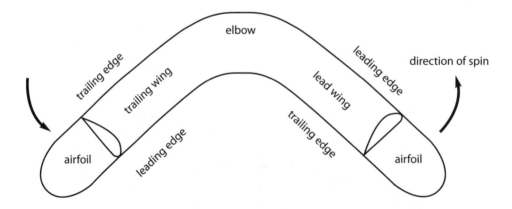

Use a rubber band to hold the two arms in a cross, or drill a ¼-inch hole in the center and insert a short piece of ¼-inch dowel to peg them together with glue. When the glue is dry, follow the throwing directions below and notice the short diameter path that the boomerang takes. To throw a boomerang, hold it vertically in your right hand, with the flat side of the boomerang to the right. Face 45 degrees to the right of any wind and fling it with lots of spin. (If the wind is blowing strong enough to stand flags out from their poles, don't throw a boomerang.) It takes patience and practice to become a good thrower.

Lefties need not be left out. They can throw a right-handed boomerang by following the same directions that righties do, but using their left hands. It will fly in a counterclockwise path, just as it does for righties. To make a true left-handed boomerang, just reverse the leading and trailing edges. Hold the lefty boomerang in your left hand, face 45 degrees to the left of any wind, and fling it with lots of spin. It will fly in a clockwise path.

Science Experiments

You can add weights (washers or pennies) to the ends of the arms to see the impact of increasing the angular momentum. You can also reshape the wings, adding drag (tape some cardboard to the boomerang and bend it upward as a spoiler) or changing the angle of attack. Make one change at a time so you can experience and understand the impact of each.

Resources

Much more information on how to make and toss boomerangs is found in a great book called *Boomerang: How to Throw, Catch, and Make It* by Benjamin Ruhe and Eric Darnell (Workman Publishing Company, 1985). For directions on how to make a cardboard boomerang for indoor throwing, see Ed's book *Fantastic Flying Fun with Science* (McGraw-Hill, 2000).

BUBBLE GUN

History of the Bubble Gun

The electric bubble gun was invented by Californian Robert DeMars; patent number 5,498,191 was awarded in 1996. DeMars has quite a few inventions to his credit, including the ever-useful clip-on handle for a beer can and a sleeping bag that's shaped like a giant teddy bear. And, along with another inventor, he came up with the resealable flip top for aluminum cans.

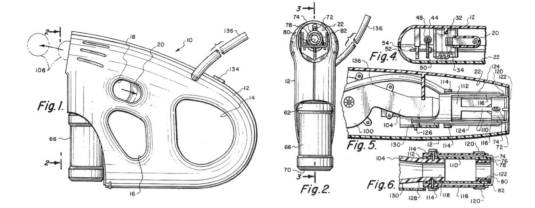

Patent no. 5,498,191

How Bubble Guns Work

Several things happen when the trigger is pulled. It lifts the applicator arm mechanically to cover the end of the nozzle with bubble juice, and it energizes the electric motor that both blows air through the nozzle and pumps bubble juice from the reservoir. The fan is a high-speed centrifugal pump; its blades spin air in a circle, forcing air outward into the duct inside the barrel. The pump, operating at a greatly reduced speed due to gearing, squeezes a plastic tube that draws bubble solution. This is known as a peristaltic pump: it squeezes the plastic to force the liquid through.

To prevent solution from spilling out of the end of the barrel when the gun is held sideways or upside down, there is a bubble juice return system. Solution collects in a funnel and flows back into the bottle past a check valve, a metal ball that can move up and down in the plastic tube.

Inside the Bubble Gun

This is a great toy to take apart. Four visible screws hold the two halves of the gun together. Two more lurk beneath the two AA batteries inside the battery case. After removing the two screws inside the battery case, reinsert the batteries so you can activate the gun while it's disassembled.

Lay it on a durable surface (bubble juice is bound to leak out), with the battery case side down. Lift off the top half. You can see that the motor drives both a centrifugal fan above and the gears below. Exhaust from the fan is forced down the plastic ducting toward the nozzle.

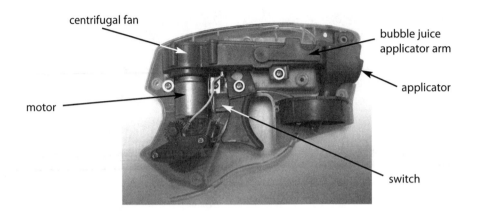

centrifugal fan

bubble juice
applicator arm

applicator

motor

switch

At the end of the nozzle is an applicator that moves up and down to spread bubble juice in front of the stream of air. This mimics how you blow bubbles the old-fashioned way: you take the wand, dip it in solution, and blow. The applicator draws a liquid sheet of solution in front of the nozzle, then air comes out.

Excess juice drains down a drip rod into a funnel, past a check valve (you can see the shiny metal ball in the tube), back into the bottle of bubble solution.

Pushing on the trigger compresses a return spring and closes a simple switch. The switch is a piece of metal (a conductor) that the trigger forces onto the metal case of the motor, thus completing the circuit and energizing the motor. The trigger also lifts the applicator at the end of the nozzle.

pump
housing

trigger return
spring

pump "impellers"
on bottom of
large gear

The shaft coming out of the bottom of the motor has a worm gear. This allows the motion to change directions 90 degrees and engage a set of gears that slows the motion and increases its torque. The end of this gearing is the peristaltic pump. Unscrew two screws that hold the gears to the gun body so you can remove the gears. Lift the bottom gear, and on the reverse side, you'll see two plastic nubs. These press against the clear plastic tubing. As they spin, they pump bubble juice along the tube from the reservoir to the barrel.

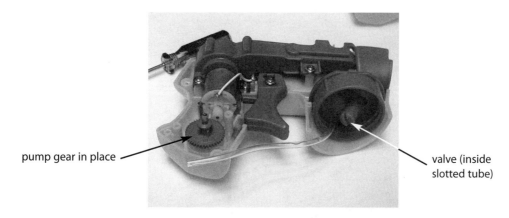

pump gear in place

valve (inside slotted tube)

Where else can you get so much entertainment for 10 bucks?

Build Your Own

We haven't built one (yet), but have seen some simple nifty bubble machines. One design incorporates an electric motor to rotate a collection of five or six bubble wands. As the motor rotates, each wand is dipped into a vat of solution. Behind the motor assembly is a small fan. Voilà—a continuous stream of bubbles!

You could also create a bubble gun by using a small gear-head DC motor to turn the wands, or you could gear down a regular DC motor. Or you could buy any of the several kits offered online.

DIE-CAST TOY

History of the Die-Cast Toy

Die-casting of toys started in the early 20th century in the United Kingdom and United States. With time and experimentation, the metal used by manufacturers became stronger and less brittle, allowing the toys to last longer.

Dinky Toys was one of the earliest die-cast toy manufacturers. After World War II, an industrial product company, Lesney Products & Co., Ltd., started making Matchbox cars and trucks, which soon became prized collectables. In the late 1960s Mattel launched its very successful Hot Wheels line. Less popular today, die-cast cars and trucks remain on the market primarily for collectors and as promotional products.

How Die-Cast Toys Work

Using scales from 1:12 to 1:76, model makers create models and use these to make molds of steel. A lubricant is sprayed into the mold to make it easier to remove the cast toy. Molten metal—an alloy of zinc, aluminum, magnesium, and copper—is forced under high pressure into the steel molds so it fills all the cavities. The mold is cooled with water to speed the process, while pressure is maintained on the molten metal. When the metal has solidified, the mold is released and the toy removed. It can be painted, and other parts, such as wheels, can be added.

Today plastics (injection molding) have replaced metal die-casts in many toys. The low cost of plastics makes them less expensive overall. There are several types of thermoplastics (high-molecular-weight polymer plastics that can be melted and frozen) including nylon and ABS, which is used in making Lego bricks. The plastic is delivered in pellet form. It is heated and forced under pressure into a closed mold that is made of steel or aluminum. Water cools the mold and plastic. The mold is opened, and an ejector pin forces the solid plastic piece out of the mold. Although we may think of injection molding as a new process, it was patented in 1869 (patent number 88,633) by inventor John W. Hyatt.

Build Your Own

As kids we used to make lead fishing weights by pouring molten lead into impressions we made in the ground, then adding a hook at the top. The lead came from old telephone cables, which were apparently used to weight the otherwise lightweight cables. A safer process would be to make plaster of paris or glue and sawdust mixtures and pour them into molds that you make in wet sand or moist earth.

> *"Way back in the mid-1900s, when I was eight or nine or somewhere thereabout, our imaginary town and farms—my brother's, cousin's, and mine—all fit neatly into the corner of our yard, where the foliage of a giant elm and a sturdy cedar blocked out the sun and sand triumphed over grass. There, sometimes for hours and sometimes for days at a time, imagination, pretend, and creativity reigned supreme. We called it play, but it was also growth and learning. Our die-cast cars (and trucks), in various sizes and stages of wear, all came from the five-and-dime store. The carefully groomed streets and roads they traversed were thoroughfares we made with hoes and rakes borrowed from our parents' tool shed. The houses and stores to which we 'drove' were piles of fallen twigs or roots that poked through the ground in weird shapes. We didn't know it, but as we built our miniature world and 'worked' and 'lived' the stories we made up about it, we were getting ready for the real world of adults, through role-playing, sharing, negotiating, and problem-solving. We were also having fun playing with toys bought and toys found."*
>
> —G. Rollie Adams, President and CEO, Strong National Museum
> of Play, home of the National Toy Hall of Fame

DIVING SUBMARINE

History of the Diving Submarine

Two brothers who owned a cosmetics laboratory first hit upon the idea of using the gas bubbles from baking powder to raise and lower toy submarines. The brothers, Benjamin and Henry Hirsch, sold their toy idea to a cereal company, which manufactured the submarines to use as premiums. It was a very popular offering. In a few months' time, more than a million diving submarines had been sent to bathtub duty throughout the United States.

Encouraged by the success of their discovery, the Hirsch brothers focused on inventing other toys and enjoyed several successes. In addition to diving submarines, they made diving scuba divers and the Magic Moon Garden.

How Diving Submarines Work

The sub shown here does repeated dives before coming to a rest on the bottom. Before launching it on its vertical mission, you pack it full of baking powder. Baking powder is a combination of an acid, a base, and filler. Once wet, the ingredients react to create carbon dioxide gas.

The gas forms slowly because it takes time for water to reach all of the baking powder. As a bubble forms, it gets caught beneath the sub, displacing some water in the process. With positive buoyancy created by the less dense gas, the sub rises. Once at the

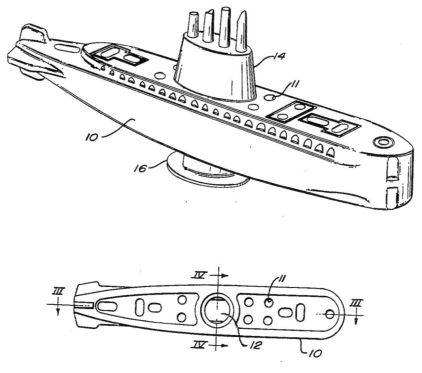

Patent no. 2,712,710

surface, the sub "burps" and the bubble escapes to one side or the other. Now deprived of its bubble of buoyancy, the submarine sinks again. The sub will rise and sink until the powder is exhausted.

Inside the Diving Submarine

Until you add baking powder to the chamber, there's nothing inside. Lift off the conning tower and add a pinch of the white powder. Replace the conning tower and you're ready to dive.

The shape of the underside of the sub is interesting. The part under the baking powder chamber is a dome that is used to accumulate carbon dioxide gas. The rest of the sub is slotted or has other holes to ensure that air bubbles don't get trapped and prevent the sub from sinking.

Build Your Own

You can make a diving submarine using a film canister and some weights. To prevent the canister from turning, hang some weights—metal washers on string—from the bottom.

Poke holes in the bottom and sides of the canister to admit water. Add a few pennies for ballast and drop in a fizzing antacid tablet. The tricks are to get the holes in the right position and to get the right weight.

The version shown below features an outer hull made of an inverted plastic cup. Instead of being connected directly to the canister with string, ballast weights are attached to wire legs on three sides of the outer hull. This design was created by Kevin Hardy at Scripps Institution of Oceanography.

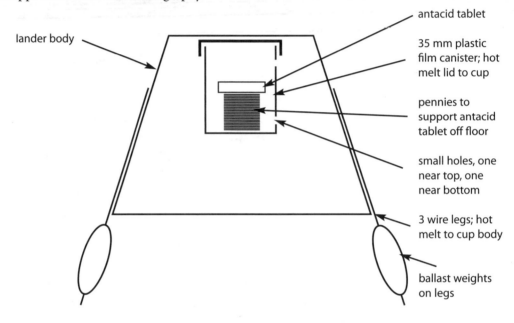

lander body

antacid tablet

35 mm plastic film canister; hot melt lid to cup

pennies to support antacid tablet off floor

small holes, one near top, one near bottom

3 wire legs; hot melt to cup body

ballast weights on legs

For super demonstrations, you can make a diving submarine from an unopened ketchup packet. Clip a paper clip onto the packet so it is slightly positively buoyant—so it floats, but just barely. Jam it into a two-liter soda bottle filled to the top with water. Screw on the lid. When you squeeze the bottle the ketchup sub should sink, and when you release the bottle it should rise. If the packet doesn't sink, add more weight. Usually an additional half paper clip will do the trick. If you do this trick subtly, people may not see you squeeze the bottle and will be amazed that the submarine rises and falls on its own.

Ask a volunteer to move his or her hand up and down beside the bottle. As the person's hand goes down, squeeze, and it'll appear that the packet is following the hand. Most are amazed at this magical performance.

Resources

A full description of this and other science demonstrations is available online at WikiDemos (http://scidaho.org/wikidemo/).

DUNKING BIRD

History of the Dunking Bird

Miles Sullivan created the Dunking Bird (also called the Drinking Bird and the Dipping Bird) in 1946. He was an engineer and tinkerer who came up with the idea for this heat engine. The patent office awarded him patent number 2,402,463 that year.

How Dunking Birds Work

If ever there was a perpetual motion machine, this must be it. The bird moves apparently without a source of energy. But it turns out that it's a great example of a heat engine. You supply the heat from a lamp or the sun.

This heat engine operates using the temperature difference between the bird's head and its body. It doesn't operate at cool temperatures—if yours is sluggish, move a hot cup of coffee or a lamp next to it to get it going.

To start the bird, pull its head down and dunk its beak in a glass of water. As water evaporates from the beak it cools the methylene chloride vapor inside the bird's head, thereby lowering its internal pressure. With reduced vapor pressure in the head and

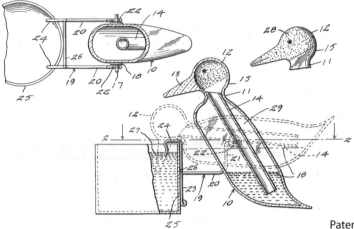

Patent no. 2,402,463

increased vapor pressure (due to the heat of the room, a lamp, or a nearby hot cup of coffee) in the base, the fluid is forced up the tube. As the fluid rises, it tips the bird's head into the glass of water. At this position the tube is tilted far enough to allow vapor to move from the head to the base, equalizing pressure. The liquid falls back to the base, rocking the bird and starting the cycle over.

Heat from a lamp or the sun causes the liquid to rise up the glass tube and the bird to tip.

The foam beak gets wet in the glass of water. Water evaporates from the beak, cooling the liquid.

This heat engine has its hot side—the base—and its cool side—the head. Take away either the heat source for the base or the cooling mechanism at the head (the water needed for evaporative cooling) and the bird will stop.

Inside the Dunking Bird

The bird's hat covers the end of the glass tube. Since the tube ends in a sharp point, we don't recommend removing the hat. The beak is made of foam that gives form to the red material that is stretched over it. Inside the glass is methylene chloride, a liquid that boils at 104° F, about the temperature of a hot tub.

ELECTRIC TRAIN

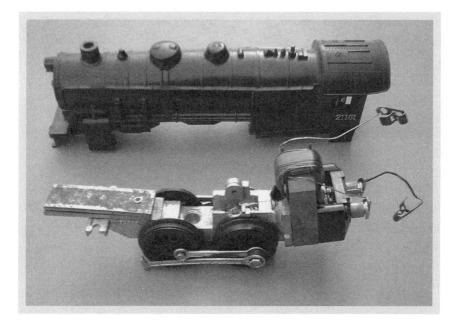

History of the Electric Train

Joshua Lionel Cowen is credited with inventing electric trains in the United States. But there were other inventors before him. When Cowen was a child he probably played with toy trains pulled by string or powered by windup clockworks, which were invented in the 1850s by George Brown. These early toys didn't run on track; that feature wasn't added until 1901.

The first electric train was created as a demonstration of the feasibility of using electricity to power real trains. Thomas Davenport powered his crude model train in 1835 with a wet cell battery. Davenport invented several types of electric motors and won the first U.S. patent for an electric motor in 1837. Although his train worked, his contemporaries weren't as visionary as he was, and his idea of electric trains died.

Märklin, a German toy company, produced an electric toy train around 1891. By 1896 an American firm, Georges Carette & Co., was selling electric toy trains. Cowen didn't make his first electric train until 1901, and he didn't intend that train to be a toy. He rented it to store owners to put in their display windows so they could capture shopper's atten-

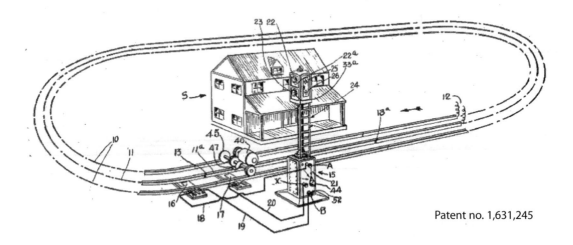

Patent no. 1,631,245

tion. However, when shoppers wanted not just to admire the train but to buy it, Cowen went into production. He used his middle name, Lionel, for his trains and his company.

Cowen credited himself as the inventor of the dry cell battery and the flashlight, neither of which he actually invented. However, one of his early employees, Conrad Hubert, had great success with one of Cowen's ideas. Hubert started making flashlights and then batteries under the company name of American Electrical Novelty & Manufacturing Company, which was later renamed American Ever-Ready Company and later still, Eveready. Hubert became a millionaire.

Lionel's major electric train competitor originated with W. O. Coleman and William Hafner, who started manufacturing mechanical toy trains in 1906. By 1910 it had become the focus of their manufacturing company, which was renamed American Flyer. However, the company languished until it was bought by A. C. Gilbert in 1938. He invested heavily to improve the design and manufacturing process and succeeded in making American Flyer a major force in the electric train business.

How Electric Trains Work

The basic idea hasn't changed since Thomas Davenport made the first electric train: metal rails are used to conduct electricity to the motor housed in the locomotive. Lionel adopted a three-rail system, with the center rail carrying the voltage and the outer rails being ground. American Flyer and others used a two-rail system, where one rail is ground and the other is charged.

Transformers vary the voltage so you can control the speed of the train. To speed up Ed's old American Flyer, he twisted a lever on a transformer that sent a higher voltage to the track.

Alternating current motors spin because the rotor is attracted to a constantly changing magnetic force. The stator winding (in the stationary part of the motor) becomes an electromagnet, changing fields 120 times per second. The tracks deliver the 60-cycle-per-second alternating current (AC) to the stator. Since opposite charges attract, the trick in making a motor is to change the current flowing through the windings of the rotor that spins inside. That way, the stator is always attracting one side of the rotor and repelling the other to make it spin.

Changing the current to the rotor windings (another electromagnet) is the job of the brushes and commutator. Pushed by springs or spring action of metal contacts, the two brushes press against two of the three sections of the commutator face. As the rotor spins, the commutator picks up different-polarity current. The current changes the polarity of the magnetic field in the rotor so that the changing field of the stator always attracts it and it keeps rotating.

Modern electric trains have the option of radio control. In these, the tracks stay at a constant voltage (about 18 volts), and the circuit boards in the locomotives regulate how much voltage the motors receive. You send radio signals to the circuit board with a handheld transmitter. Standing anywhere in the room, you can speed up, slow down, or back up the train. You can turn on its lights and clang the bell. One radio controller can control several trains at once. This isn't your father's train set—nor is it as inexpensive as his was.

Inside an American Flyer Train

The photo at the beginning of the chapter shows a 1948 vintage train. One thing we noticed right away about this train is that all the screws are flat heads. Any similar device today would have Phillips screws. Invented in 1935, Phillips screws might not have been common when this model was designed, but electric train experts tell us that American Flyer never switched to Phillips screws.

fill tube for smoke fluid

bellows

electric motor

worm gear

Although most of the action is up front in the locomotive, we first looked at the tender car. The fill cap on top and four (flat head) screws underneath hold the cover on the chassis. By removing the cover, you'll reveal a nifty electric motor that drives a bellows through a worm gear to blow smoke. Smoke fluid, a light oil, was squeezed into the fill pipe on the tender. A heating element inside heated the liquid into a smoky vapor, and the bellows pumped it through a tube to the smokestack at the front of the locomotive.

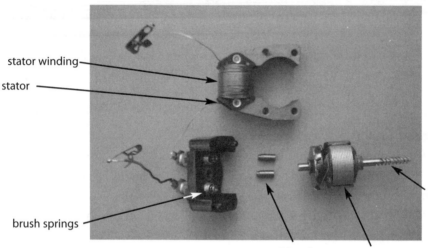

stator winding
stator
brush springs
worm gear
2 brushes
rotor

The electric motor that drives the train is located at the rear of the last set of driving wheels. The device on top, at the front of the wheel assembly, lets you reverse the engine. Called the E-Unit, this is a solenoid that rotates a drum with fingers that change polarity of the current going to the motor. When the E-Unit was engaged by the lever on top of the locomotive, the train would reverse direction whenever power was turned on. Lifting the transformer handle would cut power. Lowering it would cause the solenoid to move. Raising and lowering it once more would send the train off in the opposite direction. The process was a bit cumbersome, but that was part of the magic of skillfully operating an electric train.

We dug a little deeper inside a 1960s-era American Flyer. Like the earlier model, the electric motor is mounted to the rear of and above the locomotive drive wheels. The motor is a "shaded-pole motor" that has only one stator winding. The winding is the bright wire.

Two long bolts hold the motor together; as soon as they are removed, the brushes (two small metal rods) fly out.

Springs in the housing on the left push the brushes, allowing them to make contact with the commutator (on the right). The commutator is connected to the rotor that spins inside the stator (top). It is attached to the worm gear that turns the wheels. The worm gear changes the direction of rotation 90 degrees; the motor axis is perpendicular to the axle it's driving and the worm gear connects them. It would be easier to orient the motor parallel to the axles, but the narrow locomotive can't accommodate the motor in that position. In the model we examined, wheels on the tender car pick up the electrical power and wires carry it to the motor inside the locomotive.

Modern trains don't look like that. This recent-model Lionel locomotive has two motors, one at each end. On top of each motor is a flywheel to smooth out the speed. Between the motors is a bank of circuit boards. This train runs on radio controls.

Resources

All Aboard! by Ron Hollander (Workman Publishing Company, 2000) chronicles the history of Joshua Cowen and Lionel Trains.

ERECTOR SET

History of the Erector Set

A. C. Gilbert, the inventor of the Erector Set, was a fascinating man. Amid some controversy, he won an Olympic gold medal in the pole vault in 1908; he earned a medical degree from Yale; and he started one of the most successful toy companies in the United States. During World War I, when the government was planning to convert all toy manufacturing into manufacturing for the war effort, he "saved Christmas" for kids by testifying before the Council of National Defense on the importance of toys to learning. As soon as the U.S. secretaries of war, commerce, and the U.S. Navy crawled on the floor to play with the toys he'd brought, he'd won his argument.

Inspiration for the Erector Set came to Gilbert on a train ride between New Haven, Connecticut, and New York City. Adjacent to the tracks Gilbert saw workers erecting steel girders to carry electrical power lines. The girders inspired him to make a construction

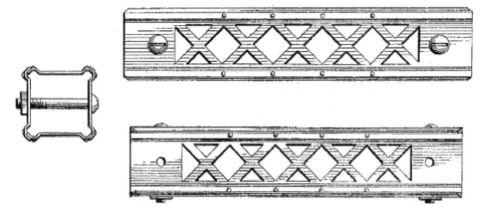

Patent no. D48,859

toy. He and his wife cut shapes out of cardboard until he found the shapes that allow for versatile construction. The Erector Set made him rich and famous. His company also made American Flyer trains, chemistry sets, and hundreds of other toys.

Gilbert's Erector Set inspired generations of (mostly) boys to pursue the study of engineering and science. Few people have had as great an impact on the careers of so many as did A. C. Gilbert. The A. C. Gilbert's Discovery Village in Salem, Oregon, (a museum that Ed once directed) has a great collection of A. C. Gilbert Company toys.

How Erector Sets Work

Structures are created by fastening the girders together with tiny nuts and bolts. Anyone who plays with an Erector Set quickly learns the importance of triangles in keeping rectangular structures from becoming parallelograms. Larger Erector Sets also included wheels and motors, which could be used to make cranes and elevators. The motors were powered by alternating current from a wall outlet, so motorized vehicles made from Erector Sets couldn't travel very far.

Erector Sets continue to be sold today, but the brand has been bought and sold several times and the current products are no longer the open-ended construction sets of Ed's youth.

Resources

The Man Who Lives in Paradise (Rinehart and Company, 1954) is the autobiography (with Marshall McClintock) of the amazing A. C. Gilbert.

ETCH A SKETCH

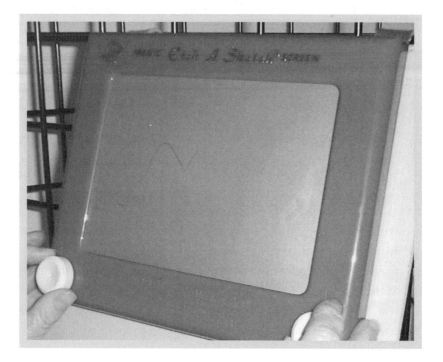

History of the Etch A Sketch

It took two tries before the Ohio Art Company agreed to purchase the toy from inventor Arthur Granjean in 1959. The French mechanic had originally called his drawing toy *L'Ecran Magique*—the Magic Screen. It was by far the most expensive toy Ohio Art had considered acquiring, and company executives initially balked at the asking price. But given a second chance to buy it, they made the purchase, and the Etch A Sketch made their company an American icon.

Etch A Sketch was patented in the United States (patent number 3,760,505), but the patent has since expired and other companies now make similar toys. Ohio Art has sold more than 100 million Etch A Sketches.

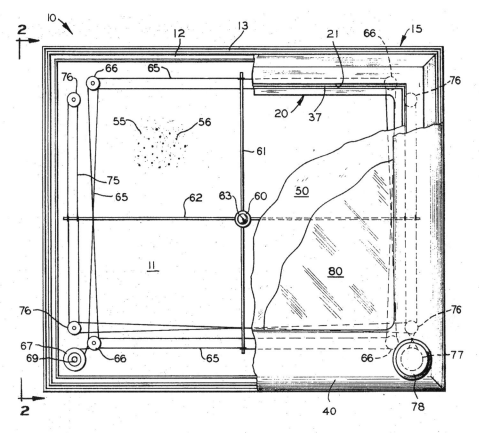

Patent no. 3,760,505

How Etch A Sketches Work

The left knob moves the pointer to the left and right. The right knob moves it up and down. The arrangement of monofilament string and pulleys controls the cursor with the pointer that touches the underside of the glass plate. As the pointer moves, it scrapes aluminum powder away from the underside of the glass, leaving a dark trail. (The fine powder sticks to *everything*, so if you open the case be prepared to spend some time cleaning everything around you.) Shaking the toy redistributes the powder (and small beads that help distribute the powder) so it adheres to the glass, erasing the image.

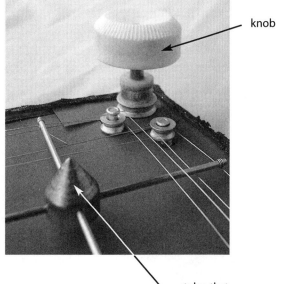

knob

stylus that scrapes powder off the screen

Inside the Etch A Sketch

This is one toy you want to avoid taking apart. It won't go back together, the aluminum powder is probably not healthy stuff to inhale, and you'll get it everywhere. How do we know you'll get it everywhere? Because we did.

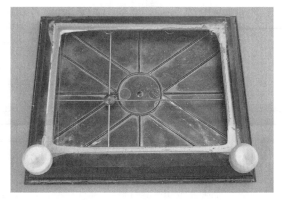

Etch A Sketch with plastic cover and aluminum powder removed

Here's how we wrestled the toy open. It wasn't pretty. We cut it open along the seam using a hacksaw (the first time we opened one) and a Dremel rotary cutting tool (each of several times afterwards). Of course as we cut through the plastic case, more and more of the obnoxious aluminum powder spilled out.

Once we had the two parts separated, more or less along the lines at which they'd been joined, we could see the details of how the stylus works.

For a less messy look inside, just use the Etch A Sketch in the normal way to scrape all the aluminum powder off the screen—great while watching a sporting event on the tube. With enough of the powder gone, the screen will be clear enough to allow you to see inside.

Resources

Kid Stuff: Great Toys from Our Childhood by David Hoffman (Chronicle Books, 1996) has more information on the development of Etch A Sketch and 40 other toys.

FRICTION CAR

History of the Friction Car

These simplest of powered toy cars have been popular for well over a century. Israel Donald Boyer and Edith E. L. Boyer of Dayton, Ohio, invented the friction, or inertia-wheel, toy in 1897 (patent number 593,174). Their design shows four wheels, with the inertia wheel suspended in the center of the "truck." The Boyers' inertia wheel consisted of two heavy cast iron wheels fixed to a spindle. To spin the inertia wheel, the user pulled on a string wrapped around the spindle. He or she inserted the spinning inertia wheel motor into the truck, and removed it when the truck came to a stop. The Boyers had previously patented a double-wheeled top, and their truck was an application of that design. They saw it as an improvement over the spring-driven toy cars of the day.

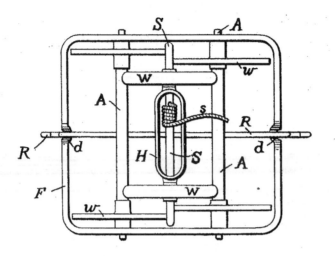

Patent no. 593,174

How Friction Cars Work

The flywheel stores rotational energy. The bigger it is and the faster it's rotating, the more energy it stores. You spin the flywheel up to speed, and then it provides power to the wheels. To get the most energy out of it, the inertia wheel is made as heavy as possible, with as much mass located as far from the center as possible. It's also spun as fast as

possible. Today's friction toys have gear systems that increase the speed of rotation from the drive wheels to the inertia wheel. A few arm-length pushes of the car get the inertia wheel spinning quickly. When released, the inertia wheel turns the drive wheels slowly.

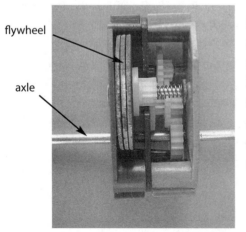

flywheel

axle

The flywheel is an important component in many machines, including the automobile starter. (There's one in the Air Hogs car we took apart, too—see page 4.) Some people have suggested that cars or buses could run on flywheel power to cut pollution in cities—that a vehicle could be powered by revving up its flywheel with an electric motor and using the spinning flywheel's energy.

Inside the Friction Car

The most interesting parts of a friction car are attached to the rear axle. Remove the car body from the chassis. You may have to yank off the wheels; they are probably held on the shaft by friction alone, so a mighty tug will remove them. The flywheel and gears are usually encased in a plastic housing that you must remove.

The flywheel is the metal disc or discs. The larger the diameter and the heavier the disc is, the more energy it can store. Twist the axle and watch the flywheel turn. In the model we inspected, every rotation of the axle turned the flywheel 16 times. By running the car across the floor, you can really get the flywheel flying! When the car is released, it moves slowly, since the gears from the flywheel increase the torque and reduce the speed.

FRISBEE

History of the Frisbee

In the early 20th century, Yale students made a sport of flinging empty pie pans across campus. The pie pans were marked with the name of the baker, the Frisbie Baking Company of Bridgeport, Connecticut. To alert someone that a metal pie pan was headed his or her way, students would call out the name printed on the pie pan: "Frisbie!" At least, that's what Frisbee lore tells us. Whether it is more fiction than fact we don't know, but it does make for a good story.

The Frisbee story jumps to 1948, when inventor Fred Morrison became interested in finding out what he could make out of the new materials called "plastics." He invented the flying disc and sold it to the toy company Wham-O. After introducing the toy in 1957 as the Pluto Platter, Wham-O learned about the flying pie pan tradition at Yale and renamed the discs Frisbees (with the intentional misspelling).

Since then, over 300 million Frisbees have been sold. The toy has spawned over a dozen games, including disc golf, Ultimate, and Guts. Over 90 percent of Americans say

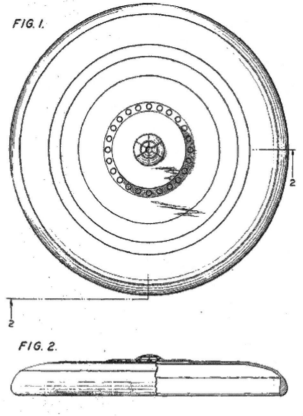

FIG. I.

FIG. 2.

Patent no. D183,626

they have played with a Frisbee—a larger percentage, we bet, than have played with nearly any other toy.

How Frisbees Work

To help understand how a Frisbee flies, try making a flying disc out of a piece of cardboard. Cut a circle about 10 to 14 inches in diameter from a corrugated cardboard box. Hold it level with two hands, then "push" it forward and release it. As soon as the disc is released into the air, the front edge rises and the disc crashes. It has too much lift, and the lift is way too far in front of the center of gravity.

Now hold the disc in your right hand, with your thumb on top and your fingers beneath. Fling it forward, giving it some spin. The disc still crashes, but notice that this time the right edge rises and flips over to the left. Try again with your left hand—the disc will crash in the opposite direction. Obviously, spin is important to determining how the disc will fly.

Forces imposed on spinning objects show up 90 degrees away (downstream) from the force. That's called "precession." When riding your bike with no hands, you can make a right turn by leaning slightly to the right. The bike doesn't fall to the right; instead, the spinning front wheel *turns* to the right. You applied force to the *top* of the wheel when you leaned, and in response the *front* of the wheel—90 degrees downstream—turns to the right.

In the same way, when a disc flies the leading edge encounters air on the underside, gains lift, and flips up. When the disc is spinning it does the same thing, but the upward

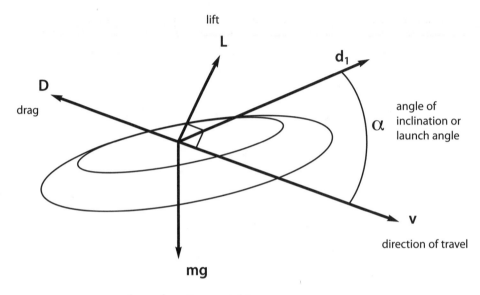

force of gravity or weight

motion shows up 90 degrees later in the spin. For a right-hand throw the lift shows up 90 degrees downstream in the clockwise direction, or on the right side. The right side lifts. Lefties see the left side rise.

Thus, a Frisbee is designed to reduce or counterbalance the lift on the leading edge of the disc. Frisbees kill some of the lift by having a blunt leading edge. This creates turbulence and reduces lift. The problem is that if you kill too much lift, you can't throw the flying disc far.

Another way to think about this is that the center of lift has to be near the center of gravity, which is the geometric center of the disc. Unbalanced forces of lift (upward on the leading edge) and gravity (downward on the center of the disc) cause a torque on the spinning disc that results in it flipping. Without a blunt edge (or a lip like the one found on an Aerobie—see page 1) or a lot of weight to counteract the lift, the edge will rise and crash.

The Frisbee creates lift by flying at an inclined angle. It deflects air downward, so the equal and opposite reaction (the air's impact on the disc) is to push the Frisbee upward. This is the force called aerodynamic lift, marked L in the diagram above. Lift opposes gravity in the balance of forces. Drag (air resistance) opposes the forward motion of the Frisbee and slows it down. As the Frisbee slows, it generates less lift and gravity pulls it to the ground. Both lift and drag depend on the angle at which the Frisbee is thrown.

Make Your Own

Cut a circle out of corrugated cardboard to make a flying disc. The circle can be any size, but a 10- to 12-inch diameter works well. You might choose instead just to use paper dinner plates or aluminum pie plates. Be creative in coming up with designs to try. A flying disc doesn't even need to be round—six-sided discs fly about as well as circular ones do.

Here are a few things you can do to make a homemade disc fly better: affix weights (washers or pennies) evenly around the outer edge; mount a second disc of the same size or smaller on top of the first one; turn one paper plate inside out and tape it to a stack of three other paper plates.

Resources

For directions on how to make other flying toys, see Ed's book *Fantastic Flying Fun with Science* (McGraw-Hill, 2000).

For an in-depth analysis of a Frisbee's flight dynamics, see Ralph D. Lorenz's book *Spinning Flight* (Springer, 2006).

FURBY

History of the Furby

Thomas Edison was the first person to add a voice to a doll. Charles Batchelor, who worked for Edison, suggested the idea as a new way to use Edison's favorite invention, the phonograph. Edison devoted significant resources to making his talking doll (which repeated "Mary Had a Little Lamb" and other nursery rhymes), but he discovered that the delicate mechanisms couldn't stand up to the rigors of shipping. The toy business was a bust for Edison.

A century later, another pair of independent toy inventors decided to make a "smart" and cuddly toy using solid state electronics (that survive shipping very well). Freelance designers Caleb Chung and Dave Hampton created the Furby, which can do far more than simply repeat nursery rhymes. It has a vocabulary of 200 words and can converse with its owner or another Furby; its ears, eyes, eyelids, and mouth move; and it knows when its tummy is being rubbed and when the lights are turned out.

Within 12 months of Furby's 1998 release, 27 million units had been sold.

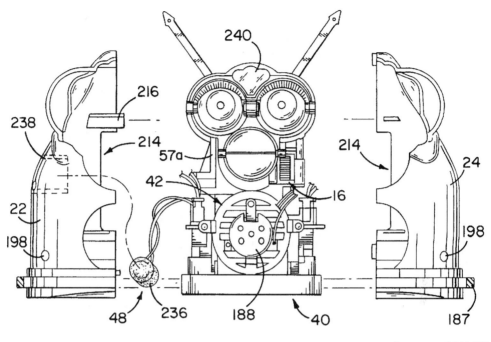

Patent no. 6,149,490

How Furbys Work

When the Furby was first released, a myth was also born: that it would learn to speak by listening to what people said and repeating, in its synthesized voice, what it had heard. This was such a concern that the United States National Security Agency banned Furbys from its premises. In reality, the Furby's entire vocabulary was preprogrammed, but it was designed to vocalize more and more words over time, so people assumed it was learning from them. It was quite a clever trick, really.

All of a Furby's movements are activated by a single DC electric motor. A circuit board controls the motor, the sensors, and the sound circuit.

Inside the Furby

Getting inside this toy is a job. The first task is to get the skin off. It's held on in several places by stitching, which we cut with Swiss Army Knife scissors. Also holding it on is a cable tie that holds the edge of the fur around the base.

We cut the ear coverings away from the fur and pulled them off. Inside each ear is a plastic gear and lever that support and move the ears.

The fur is stitched to a faceplate and we cut the threads. With the fur removed, we opened the body by removing a couple of screws.

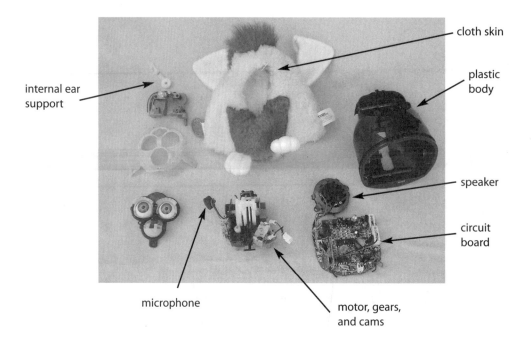

internal ear
support

cloth skin

plastic
body

speaker

circuit
board

microphone

motor, gears,
and cams

It's amazing to see that all of the Furby's actions are powered by one electric motor. The mouth, eyes, eyelids, and ears all move. And, seemingly, they move independently. That is, the eyes don't necessarily move when the ears move. Movements are controlled by the positions of cams driven by the motor. One cam in the back closes and opens a micro switch.

The motor that drives all this is located on a plastic board with the gears. To find it, look for diodes and a capacitor soldered across the motor's terminals. The motor faces downward and drives several gears, which drive a worm gear that changes the direction of rotation from a horizontal plane to a vertical plane.

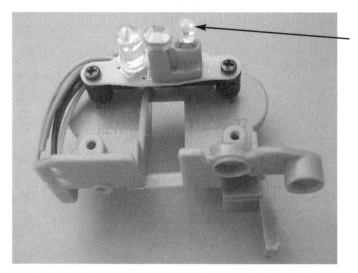

light sensor,
LED, and IR LED

Circuit Bending

Although the Furby is no longer as popular as it once was, it is still very important to a group of hackers and tinkers who enjoy "circuit bending." This is the process of rewiring or creatively short-circuiting electronic toys and instruments to create new sounds and noises. These sounds are often used in experimental music and as sound effects for movies and videos. If you've had all the fun you want with your Furby and would like to see what else you can get it to "learn," you could try some circuit bending—but remember, it's extremely difficult to "unbend" a circuit.

An introduction to circuit bending a Furby can be found at the Web site Circuit Bent (www.circuit-bent.net/furby-bending-tutorial.htm).

In the middle of the Furby's chest is a speaker. A microphone is located on its right side. Loud noises "startle" the Furby and awaken it from a resting state. Above its eyes are a light sensor, LED, and infrared sensor. When the light sensor reads no light for more than 15 seconds, it puts the Furby into a "sleep" state. The LED operates in the infrared (IR) range and communicates to other Furbys. It works much the same way as a TV remote control, which also uses IR. The infrared sensor receives information from other Furbys.

The circuit board is crammed with components, including an inversion switch. If the Furby's operation runs amok, you restart it by turning the toy upside down. After removing the board, flip it upside down to hear a weight in the inversion switch move.

There's a ton of technology inside a Furby. The most amazing thing is that you can buy all of it, crammed into one cute little package, for just a few bucks.

"My friend (Dave Hampton) and I wanted to create a little creature that could be a friend. We both had different skills, so we collaborated. Dave worked on the electronics and the brain and I worked on the body and the mechanics. We wanted it to be small, cute and funny, and not too expensive so everyone could have one."

—Caleb Chung, cocreator of the Furby

Resources

Edison: Inventing the Century by Neil Baldwin (Hyperion Press, 1995) tells the story of Edison's venture into toy manufacturing.

GYROSCOPE

History of the Gyroscope

French scientist Jean Bernard Léon Foucault invented many things, and science fans everywhere remember him for the Foucault pendulum, which demonstrates that the earth rotates on its axis. Foucault pendulums grace the lobbies of science museums worldwide. But Foucault is also known as the inventor of another marvel. Although the gyroscopic effect was first reported by Johann von Bohnenberger in 1817, it was Foucault who, in 1852, created the first actual gyroscope.

The gyroscope made it to the big leagues of technology with Elmer Sperry's development of the gyrocompass. For decades, the gyrocompass was used to help navigate ships, planes, and rockets.

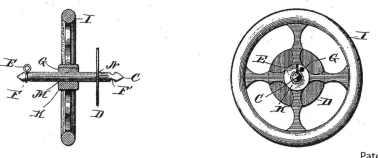

Patent no. 461,948

James Wilson is the first person to have received a U.S. patent (patent number 461,948) for a toy gyroscope. Toy gyroscopes started spinning under Christmas trees around 1917.

How Gyroscopes Work

You can learn how a gyroscope works via words or mathematical symbols—or by playing with one. The latter method is preferable, at least for the initial stage of study. Wind a string around a gyroscope's shaft and pull smartly to get the wheel spinning. The wheel is weighted around its rim so that it can attain high angular momentum. The more mass in the outer rim of the wheel and the faster it spins, the more angular momentum it has.

If you balance one end of a spinning gyroscope shaft on a finger, you'll see that the other end won't tip over, but will instead slowly spin around in a horizontal circle. This rotation is called precession. The downward torque (the weight multiplied by the moment arm, or the distance from the center to where your finger supports the weight) would cause the gyroscope to fall if it weren't spinning. But when it is spinning, its torque causes it to precess. For more information on precession, see the discussion of how Frisbees work (page XXX).

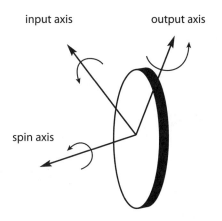

If a wheel is rotating around the spin axis, and the input axis is rotated left as shown, the wheel will turn left around the output axis.

Inside the Gyroscope

Inexpensive gyroscopes are spinning wheels that turn inside a set of perpendicular rings. The shaft of the wheel usually has a hole so you can insert a length of string, then wind the string around the shaft in order to spin the gyroscope quickly.

Build Your Own

The closest we've come to making a real gyroscope is a top. It's not quite the same, but a top provides the same types of experience. We use either 2-inch wood wheels (purchased from Woodworks Ltd., www.craftparts.com) or wheels we've cut using a circle saw. Both wheels slide onto a ¼-inch wood dowel.

First, sharpen one end of an 8- to 10-inch-long dowel in a pencil sharpener. This is the top's spindle. Then glue a wheel on the spindle about 2 inches above the sharp point.

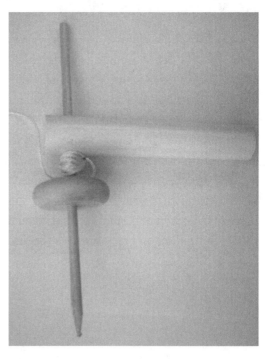

The wheel is the top's bob. (CDs also work for bobs. You can get a packet of CD inserts that fit onto ¼-inch dowels from the KELVIN Web site, www.kelvin.com.) It is interesting to try spinning the top without first gluing it together: unless it fits tightly, the spindle will turn inside the bob and the top will fall over.

Once the top is glued together, spin your top by hand. Placing the sharpened end on the floor (or on an upside-down bowl), hold the spindle between your hands. Pull one hand toward you while pushing the other hand away to impart spin.

But wait, there's more! Next, build a launcher. A short piece of ¾-inch PVC pipe works great. If you have a choice, get the low pressure, or thinner, pipe (schedule 200). Drill a 5/16-inch hole through both "sides" of the pipe, about 1 inch from the end. Clean out the burrs and you're ready to spin. Wrap a piece of string around the spindle of your top, just above the bob. Slide the spindle through the two holes in the PVC pipe. Rest the sharp end of the spindle on the ground and hold the PVC pipe horizontally, positioned just above the wound-up string. Hold the pipe stationary while pulling on the string horizontally. Once the string has come off the spindle and the top is spinning smartly, lift the launcher up and away. A good launch will last more than a minute.

A bike wheel also makes a great top. Get a set of pegs (from a bike store) that screw onto the threaded shafts of a front bicycle wheel. Hand spin the wheel to do all sorts of cool tricks.

Science Experiments

Try bobs of different sizes and weights until you find the best size and weight for maximum spin duration. For experiments in visual perception, cut out some white card stock in the shape of a circle and punch out the center so it slides onto the spindle. Tape it to the bob. Try drawing different-colored patterns, such as large circles of different colors or long colored lines, on the card stock circle to generate optical illusions.

HELICOPTER

History of the Helicopter

The ancient Chinese had a bamboo flying toy that flew like a helicopter. But it wasn't until the 19th century that a toy helicopter was patented in the United States—by Howard Tilden, in 1869. Both toys predated the 20th-century invention of a life-sized helicopter.

The miniature version played a pivotal role in the invention of the airplane. The Wright Brothers started thinking of flight when their father gave them a toy helicopter.

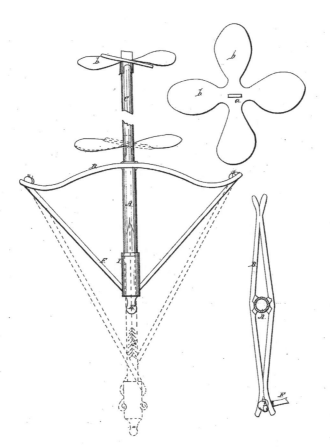

Patent no. 85,874

How Helicopters Work

The blades of a toy helicopter are shaped to generate lift as they move through the air. To get the blades moving fast enough, you spin the helicopter shaft by pulling on the ring attached to the string.

The blades are shaped to push air downward as they spin. As the blades force air down, the equal and opposite reaction (Newton's third law) is the air exerting an upward force on the blades. This upward force is the lift that supports the helicopter. Thus, the helicopter flies by blowing air down. If you have been around an operating helicopter, you know that it generates considerable downdraft. You could estimate the weight of a hovering helicopter if you could measure the total downward force of the air.

Messing around with a toy helicopter and thinking about how it generates lift is a great way to understand how airplanes fly. In a helicopter, the rotating blades are both the mechanism of propulsion and the wings that provide lift. In a plane, the wings are fixed, but they generate lift the same way: by forcing air downward.

Inside the Helicopter

The wing is a molded piece of plastic that is set in a circle for support. The center has splines (or notches) that fit into the top of the launcher.

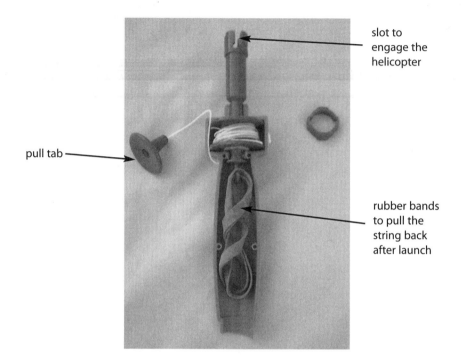

slot to engage the helicopter

pull tab

rubber bands to pull the string back after launch

The launcher is the more interesting piece. After removing the ring at the bottom, we found that the two halves came apart. The string that spins the plastic helicopter blades is wrapped around the drum inside. Pulling the string also winds up two rubber bands, which provide restoring force to wrap the string back onto the drum.

Build Your Own

There are several ways to make a helicopter-like flying toy, but making the launcher is more difficult.

A spin-to-launch helicopter requires a blade and a dowel. You can find blades (propellers) at hobby stores, or you can make your own from a thin piece of wood. Start with a piece that is ¼ inch thick, ¾ inch wide, and 8 inches long. Find the center and drill a ¼-inch hole. Using sandpaper, a rasp, or a knife, shape the two wings. Make sure that the leading edge of one rotor will follow the trailing edge of the other (in other words, make a helicopter blade, not an airplane wing). Cut a 7-inch piece of ¼-inch dowel and glue it in the hole.

To fly your creation, hold the dowel between the palms of your hands with the propeller toward the sky. Push your hands past each other to spin the stick and watch the helicopter fly. This requires some skill, so don't be discouraged if your initial flights are brief.

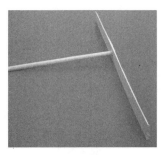

Instead of a wood rotor, you can make one out of cereal box cardboard. Cut a piece of cardboard 8 inches long and ¾ inch wide to serve as the rotors. Find the center, and using a single-hole punch, punch out a hole that is big enough to hold a drinking straw. Make cuts about ½ inch from the hole on both rotors, but on opposite sides. These cuts let you fold the rotors to make flaps. Make a second set of cuts about ¼ inch from each end of the rotor on the same sides as the first cuts.

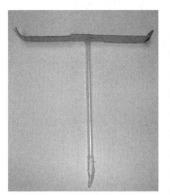

Now make two folds. The first forms the flaps on each side of the rotor. Push the flaps down at a 30- to 40-degree angle from the rest of the rotor. You can adjust these flaps to assess the impact of angle on lift. Too steep an angle adds more drag than lift.

The second set of folds adds a winglet to the end of each rotor. Bend the outer ¼ inch of each rotor either up or down, so that the winglets are perpendicular to the rotor. (On your next trip to the airport, notice how many of the newer jets have winglets at the end of their wings. The winglets prevent air from sliding off the end of the wings and reducing the lift.)

Insert a drinking straw in the center hole and hot glue it in place. A fat straw (for milk shakes) is heavier and works better than a typical straw. However, you can use regular straws—just jam a second straw or half a straw into the first one to get more weight.

If the bottom of the straw wobbles wildly and the helicopter falls, it's telling you that it wants more weight. Add more weight by jamming another straw into the bottom of the first one. Your model is too heavy if it flies without wobbling but immediately falls to the ground. Trim some weight by cutting sections off the second straw until you have the ideal weight.

Resources

Super Flyers by Neil Francis (Addison-Wesley, 1988) shows how to make several simple helicopters.

Understanding Flight by David F. Anderson and Scott Eberhardt (McGraw-Hill, 2001) gives good descriptions of lift and how planes and helicopters fly.

HULA HOOP

History of the Hula Hoop

Kids from many cultures have played with hoops, either by rolling them on the ground or swinging them around their hips. The toy that we know as the Hula Hoop is a child of the plastics age and the fertile minds at the toy manufacturer Wham-O. Richard Knerr and Arthur "Spud" Melin, founders of the innovative company, created the Hula Hoop in 1957 after an Australian friend suggested the idea. They made their hoops out of a new plastic material that had been invented just six years earlier: high-density polyethylene. Wham-O sold 100 million Hula Hoops in the first two years. Periodic promotions have rekindled the surge of interest in this toy. Today there are many new Hula Hoop variations, including hoops that fire dancers ignite and spin as part of their fire shows. Some Hula Hoops are translucent and filled with different colored light-emitting diodes (LEDs) that flash in fun patterns, making the toy more exciting to spin in the dark.

How Hula Hoops Work

Why doesn't the hoop fall to the floor? OK, it *does*, when Ed tries to spin it. But a skilled person can keep it spinning for a long time—the record is 90 hours. So how do they do it? With gravity trying to pull it down, what's holding it up?

Friction! As you swing your hips, the friction between your clothes and the hoop provides the traction to keep the hoop spinning. Part of the force you exert with your hips balances the frictional forces and drag that are slowing the hoop down as it spins in the horizontal plane. And, part of the force you exert balances out the gravitational force pulling it down. It would be impossible to successfully spin a Hula Hoop around your waist while wearing a Teflon coat, as the frictional force generated would be too small.

Not only can you learn science with a Hula Hoop, but you can also lose weight. Some estimate that you burn about 100 calories in 10 minutes of vigorous hooping.

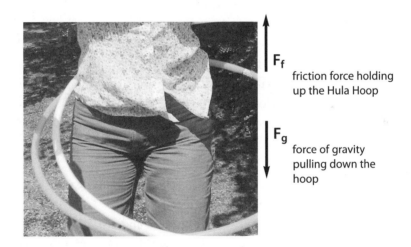

F_f friction force holding up the Hula Hoop

F_g force of gravity pulling down the hoop

In terms of energy, you provide all that is needed to both spin the hoop and keep it from falling down. Your personal gyrations provide the kinetic energy to keep the hoop going.

Inside the Hula Hoop

Most Hula Hoops have one to three steel balls that can spin around inside. They appear to be there only to produce noise, not to enhance your skill at keeping the hoop up.

Science Experiments

Try wearing different clothes while spinning the Hula Hoop to see if some fabrics have a greater frictional coefficient that makes it easier to keep the hoop up. Some synthetics may reduce friction, making the task much more difficult.

Adding weights to the hoop will make it easier to spin. The greater its moment of inertia—the weight at a distance from the center—the easier it is to spin. Try adding some fixed weights (ball bearings, bolts, pebbles) at points of equal distance from each other around the circle. So that the weights don't hit your hips while you're swinging, cut open the hoop and glue them in place inside. They'll increase the angular momentum of the spinning hoop.

By adding two weights, each at opposite sides, you can perform a neat demonstration of the moment of inertia. Hold the hoop vertically with the weights at the top and bottom (one under your hand and the other closest to the floor), then twist the hoop back and forth around a vertical axis. You'll find that this is easy to do. Now rotate the weights so that one is at 9 o'clock and the other is at 3 o'clock, and you'll find that the hoop twists much more slowly. This is because the concentration of the weight is farther from the central axis, and it therefore takes more energy to get the hoop spinning.

According to Time *magazine, Paul "Dizzy Hips" Blair has set several world records with Hula Hoops. He has run a mile while hooping at a pace of 7:47, and 10 kilometers at 1:06:35. He also held the record for spinning the largest hoop, about 13.6 feet in diameter. This record was broken two years later when Ashrita Furman spun a hoop about 14.5 feet in diameter. A woman in Australia spun 100 Hula Hoops simultaneously.*

One final demonstration shows the friction force of a floor. Hold a hoop vertically with one hand and give it a powerful backspin as you toss it forward. The hoop will slow down, then roll back to you. Its speed and delay time will vary depending on the frictional coefficient of the flooring.

Resources

See *Turning the World Inside Out and 174 Other Simple Physics Demonstrations* by Robert Ehrlich (Princeton University Press, 1990) for this and other demonstrations on the moment of inertia.

JACK-IN-THE-BOX

History of the Jack-in-the-Box

The origins of the jack-in-the-box are in question. The story we've found is pretty dark and unsuited for a children's toy: that "Jack" is slang for "slave," and that a jack-in-the-box was a slave kept in a small box. Children would provoke Jack by pushing sticks through holes in the box and would be rewarded by his cries and sudden movements. When the jack-in-the-box toy was created, the inventor wisely chose to make Jack a clown instead.

How Jack-in-the-Boxes Work

Turning the crank rotates a worm gear. This turns a drum that has a tune recorded on it as protruding pins. Each pin on the drum plucks a note on one of 18 metal tines. (By comparison, note that the record for the largest number of tines in a music box is 400.) The other side of the drum turns the gears that unlatch the top. A cam connected to the gears lifts the latch. Once the cam has passed, a spring underneath the latch pulls it back down so the lid can be closed again. Jack himself is a spring that is wrapped in cloth with a rubber or plastic clown head attached to it. When the lid opens, Jack pops out.

This toy illustrates several physical phenomena. The springs follow Hooke's law, which states that the force exerted by a spring is proportional to the length to which it's extended. The sounds are generated by vibrating metal tines. The tines vibrate at a natural frequency that is determined, in part, by their lengths. Each vibrating tine sets air molecules vibrating at its frequency so you can hear a variety of tones. The gears allow the direction of motion to change and allow the rotational rate of motion to change according to the ratio of the numbers of teeth in connecting gears.

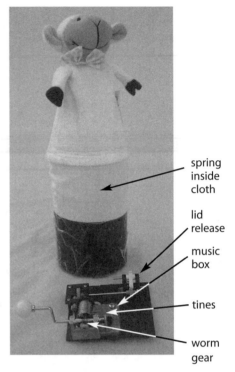

spring inside cloth

lid release

music box

tines

worm gear

Inside the Jack-in-the-Box

Some models are assembled with screws that make them an easy take-apart. The one we tinkered with, however, was made by someone who didn't intend for us to peer inside. We had to cut a hole in the bottom to pull out the spring and puppet. The spring is housed in a plastic cylinder, and it extends up to the neck of the puppet.

The winding mechanism is screwed to the inside wall and is held in place by screws on the outside of the box. To get the mechanism out, however, we had to either pull the knob off the crank or use tin snips to cut the box. We chose the latter method. Bottom line: we trashed the metal box to get the two components out.

Two screws hold the music box tines. If you remove the tines, you may find it difficult (and annoying) to get them back in place when you reassemble the toy. Line up one of the end tines with the last row of bumps on the drum, then tighten the screws to get the music box back together.

Most of today's jack-in-the-boxes use rubber belts to convey the motion of the crank to both the music box and the release mechanism. The belts have bumps on them for each note that is to be played, and these bumps lift the metal tines as the belt rotates.

KALEIDOSCOPE

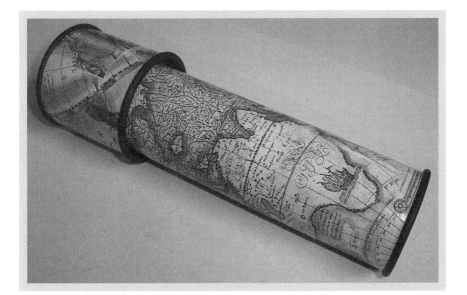

History of the Kaleidoscope

Apparently the ancient Greeks had kaleidoscopes, but like so much ancient knowledge, people forgot about them during the Middle Ages. Sir David Brewster, a Scottish scientist, reinvented the kaleidoscope in 1816. His patent describes a kaleidoscope as "an instrument for creating and exhibiting an infinite variety of beautiful forms." Brewster was a prolific scientist and writer with hundreds of papers and several important editorial positions to his credit. His toy quickly became popular throughout Britain and the United States.

Charles Bush, a Prussian immigrant with an interest in optics, promoted kaleidoscopes in the United States in the late 19th century and patented his own kaleidoscope design. Included in his patent is a design for a kaleidoscope that features a liquid-filled object box. As the kaleidoscope was turned (and for some time afterward), air bubbles would float through the liquid, making unique patterns.

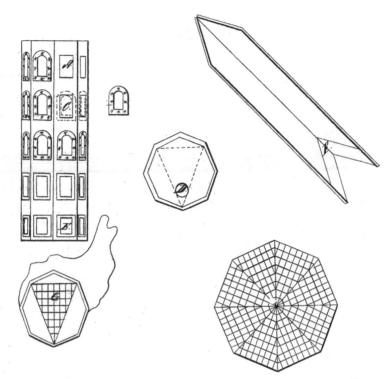

Patent no. 66,134

How Kaleidoscopes Work

There are several kaleidoscope variations, but most have either two or three mirrored surfaces inside a long cylinder. David Brewster's design had two glass mirrors set at a small
angle to each other. These generate a single, radially symmetric image. Many of the toys
you find today have three equally spaced shiny metal or metal foil mirrors set in an equi-
lateral triangle. This design generates more complex patterns. Some kaleidoscopes have
four mirrored surfaces, set in a rectangular form; as they are moved, the reflected images
move from one side of the view to the other.

Regardless of the number of mirrors inside, there are several ways in which kaleidoscopes may present images to be reflected. The least expensive model, called a teleidoscope, has a spherical lens at one end. Rather than looking at stuff inside the teleidoscope,
you look through it at objects in the space around you. Images of these objects are reflected
multiple times by the mirrors inside.

The classic kaleidoscope has a chamber located at its far end (the one most distant from
you as you peer through the device). It contains colorful materials that tumble and twirl as
you turn the chamber. Some chambers are filled with liquid and objects that float in it.

Instead of closed chambers, some kaleidoscopes have one or two translucent wheels
at the far end. You spin the wheels to change the image.

Inside the Kaleidoscope

There are two parts to a kaleidoscope: the viewing tube and the object box or wheel. Inside the tube are the reflecting surfaces. Inside the object box are plastic or paper shapes that spin and fall as you rotate the box.

Build Your Own

You can make your own small kaleido-scope using two 35mm film canisters and either three plastic microscope slides or aluminum foil. Microscope slides are available from science supply houses and some education-oriented online stores. The plastic slides are 1 inch wide and 3 inches long, almost a perfect fit inside the canister.

But first, turn your attention to the film canisters. Cut off the bottom of one canister with a coping saw. This saw cuts right through the plastic.

Cut a hole about ¼ inch in diameter in the center of the bottom of the second canister. An easy way to do this is to use a ¼-inch paddle drill bit. Hold the shaft in one hand (or in a pair of vise grip pliers) and twist it through the bottom, making sure you don't keep going into your other hand. It's best to hold the canister on a piece of wood so that when the drill bit cuts through it, the bit encounters the wood and not anything with nerves and blood.

If using foil, glue a piece onto an index card or other card stock, shiny side out, and cut out a 3-inch square. Trim ⅛ inch off one side. Make two folds to create three "mirrors" of equal width, then tape the edges together, with the aluminum foil on the *inside*. You should end up with a 3-inch-long triangular tube, with each side of the triangle being just less than 1 inch wide.

If using microscope slides, place three of them side by side on a clean table (shiny sides down—these are the "mirrors"). Tape them together with clear tape. Now pick up the slides and make a triangle out of them: pick up the outer edge of the top slide and connect it to the outer edge of the bottom slide. Tape this joint with clear tape. Now you see what's neat about this project. The three slides fit almost perfectly into the film canisters.

Push the slides or the aluminum foil tube into the canister that doesn't have a bottom. Slide the other canister on top. Bring the device up to your eye and look through the hole you drilled toward a light to see the reflected images.

Now you can get creative by adding objects inside the kaleidoscope. Take it apart and set the parts aside for reassembly. Cut out a circle of clear, rigid plastic so it just fits inside a film canister. The best place to find clear plastic is in the recycling bin, where in its prior life it protected food or some other product. (Wash and dry the recycled plastic before you use it.)

Put a translucent canister cover on the end of the currently bottomless canister, then drop a few beads, sequins, or found objects into it. Push the clear plastic circle over the objects to hold them in place. Next, reinsert the slides or the aluminum foil tube. It will hold the plastic cover in place. Reassemble the rest of the kaleidoscope. When you rotate the kaleidoscope, the beads or other objects fall around inside, changing the images.

MAGIC 8 BALL

History of the Magic 8 Ball

The history of the Magic 8 Ball is as murky as the fluid inside the toy. Apparently a Cincinnati clairvoyant inspired the fortune telling device. In the 1940s her reprobate son, Albert Carter, invented the device and went into business with Abe Bookman, who is often credited with the invention. (The company name—Alabe Crafts Company—was a combination of the first letters of Albert's and Abe's names.) A professor of psychology provided the first collection of 20 phrases the Magic 8 Ball might say. A few design changes and name changes later, and the ball went big in the 1950s. Tyco purchased rights to the Magic 8 Ball, and more recently Mattel purchased Tyco. Today about a million balls are sold each year. Only a clairvoyant could have foreseen such success.

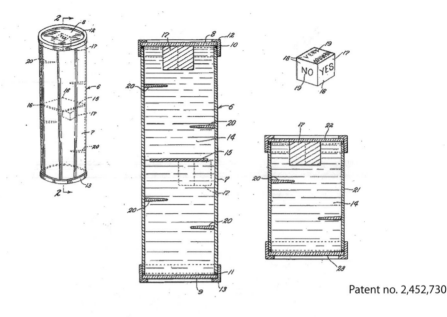

Patent no. 2,452,730

How Magic 8 Balls Work

The ball's answers are inscribed on an icosahedron (20-sided object) that floats in a reservoir of blue fluid. The icosahedron is slightly less dense than the fluid—it's hollow, and its sides have gaps that allow the object to fill with the fluid, which allows for proper buoyancy—so it slowly floats to the top of the ball, where a window allows the user to read whatever side of the icosahedron happens to appear.

When the ball is turned over the icosahedron tumbles inside the fluid-filled reservoir, rendering answers at random as the ball is turned upright again.

Invariably, however, bubbles form inside the cylinder, and a major effort has been made to keep them away from the viewing port. If they were to show up there, they would block you from reading the fortune. Over the years engineers created a bubble eliminator. To see that, you have to look inside the ball.

Inside the Magic 8 Ball

We had always thought the entire ball was filled with fluid. Not so. We used a Dremel rotary cutting tool to cut the ball in half along its equator. Inside is a cylinder that falls out as the two pieces are separated. (We were relieved not to find ourselves covered with blue fluid.)

We removed the seal on the cap of the cylinder by taking out three screws, then we drained the blue liquid, a mixture of alcohol and blue dye. The cylinder has three parts: the cap, a bubble eliminator, and the main body where the Magic 8 Ball icosahedron resides. The bubble eliminator works by collecting bubbles in the fluid and blocking their

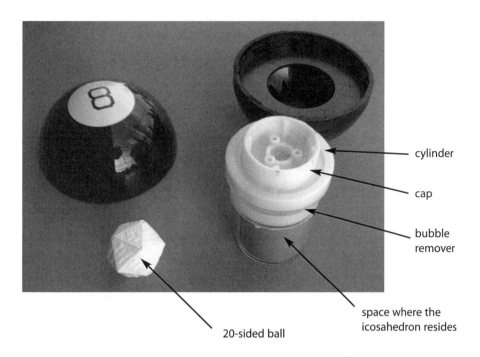

cylinder

cap

bubble
remover

space where the
icosahedron resides

20-sided ball

return to the main body. Every time you turn over the ball to get a new answer, any bubbles contained inside will float to the top and pass through the hole in the top of a funnel into the collector, where they are trapped. Shaking the ball allows the air to reenter the fluid and prevents the bubble eliminator from working effectively. We cut open the cylinder to extract both the bubble eliminator and the answer "ball."

The answer ball is an icosahedron, or 20-sided object with triangular faces. On each face is printed an answer. Half are affirmative; the rest are split evenly between negative and vague. No logic circuits or all-knowing being was found inside. Your fortune comes down to a flip of the wrist. If you don't like the answer, try again!

Resources

The beautiful book *Timeless Toys* by Tim Walsh (Andrews McMeel Publishing, 2005) features photos of a wide variety of Magic 8 Balls and provides a more detailed history of the device.

Magna Doodle

History of the Magna Doodle

When engineers Yasuzo Murata, Takeo Yokoyama, and Hiroshi Murata of the Pilot Pen Corporation tried to come up with a dust-free chalkboard in 1974, they didn't know they were about to create a very popular toy. Their patent (patent number 4,143,472) describes the invention as a display device. The company sold the toy rights to Fisher-Price, which has sold some 40 million Magna Doodles.

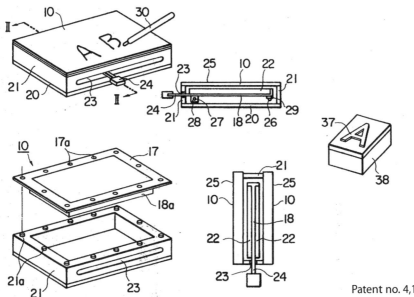

Patent no. 4,143,472

How Magna Doodles Work

The Magna Doodle is a magnetic drawing board. The magnetic "pen" pulls iron oxide particles up to the viewing surface to form images and words. The board is cleared via the magnetic "eraser" located underneath the screen, which pulls the particles down and out of sight.

Each of the cells you see in the face of the toy is closed to prevent the iron oxide particles and white liquid inside it from migrating to the next cell. To see this, lower the magnetic pen to a cell and watch it attract the black particles. The particles from one cell won't cross into the next cell. Watch, too, as you erase the images on the screen. As the eraser—a magnetic bar that

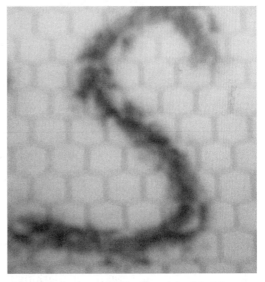

Cells contain the white liquid and the black iron filings

slides underneath the screen—moves close to a cell, it pulls the filings in that cell toward it, below the surface of the liquid, where it is no longer visible.

As the magnetic pen is moved over the screen, it attracts iron oxide particles in nearby cells to the tops of their cells, turning the area dark. The tiny filings have the same density as the liquid, so they don't fall under the force of gravity. The filings are coated with a resin to prevent them from breaking apart into such small pieces that the magnetic force can't pull them to the surface.

Inside the Magna Doodle

There are many sizes and forms of this toy. We recommend taking apart one of the larger models, as they are easy to disassemble. If you're careful, you can take apart your Magna Doodle and get it back together. With a medium-sized flat screwdriver, pry open the back. Start with the top two plastic rivets. Slide the screwdriver into the space between the two pieces (the back and front) adjacent to a rivet. Bring the screwdriver tip up, underneath the back piece, and it will pop out the rivet. Repeat for the other top rivet. Then slide the screwdriver along the sides between the two pieces and pop out those rivets.

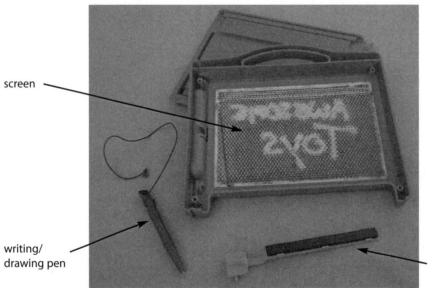

screen

writing/
drawing pen

slide eraser

When you have removed the back, you'll see that there is nothing holding in the three components: the slide eraser, the marking pen, and the screen. The slide eraser is a bar magnet connected to a piece of plastic. The plastic piece slides in a trough in the back. The marking pen is a small magnet attached to the end of the plastic "pen." (Some pens feature a wider magnet at the other end as well, which is used for broad strokes.) The pen is attached to the case by a string. That's all there is to take apart, unless you want to cut open some of the honeycomb cells. We don't recommend making that mess. (We did, and we were covered with white liquid and iron oxide pieces.)

NERF GUN

History of the NERF Gun

By the time he formed his own toy design company, inventor Reynolds Guyer had already created the game Twister for Milton Bradley, so he was familiar with the toy and game industry. Guyer started to think about toys he could create from polyurethane foam, and he ended up with a ball, which he sold to Parker Brothers (that firm was later sold to a succession of companies as the toy industry consolidated). In 1970 Parker Brothers introduced the NERF ball to the world. Since then they have added other types of NERF balls (such as footballs), and games and toys that incorporate foam balls and rockets. One such toy is the NERF gun.

How NERF Guns Work

When the piston is pumped, it forces air through three ports into the gun's three white reservoirs (one for each barrel). As the air pressure builds, it pushes the plastic inserts inside the reservoirs forward, blocking air from escaping out the barrels. Each pump of the piston forces more air into each reservoir. Oddly, though, air that pushes the NERF "darts" is downstream of the trigger. Furthermore, there isn't any direct connection from the trigger to the three reservoirs. How, then, does the trigger release the built-up pressure and fire the gun?

As the trigger is pulled, it depresses the valve in the first port, letting a blast of air escape from the first reservoir the same way it entered. As the air rushes out, the plastic insert slides back, preventing any more air from escaping this way—but at the same time opening up the air passage to the barrel. The remaining air rushes out the business end of the barrel, propelling the dart. In effect, the insert is an air-pressure-activated valve. Pretty ingenious.

Pulling the trigger farther opens the second port, releasing air from its reservoir and then out the second barrel. A final tug on the trigger releases air from the third reservoir.

Inside the NERF Gun

Remove the single screw that holds the plastic cover on the front of the gun and a similar piece on the back. With the front cover removed, you can pull off the ammo holder. Use a small screwdriver to pry the tabs loose so that the rear cover pops off. Now you can unscrew the half-dozen screws that hold the two halves together.

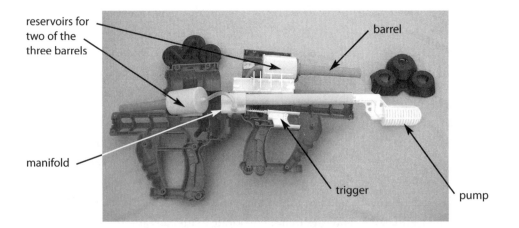

reservoirs for two of the three barrels

barrel

manifold

trigger

pump

As you separate the two halves, the trigger-return spring will fly off in some direction, so protect your eyes. Pull the three barrels off the air chambers and set them aside.

Slide out the pump handle/piston and you'll see an O-ring at its end. The O-ring forms a seal between the piston and the cylinder walls so that air is compressed into the three reservoirs.

If you're following along at home with a working NERF gun, this is the time to put down the tools, as the next steps will transform the toy from functional to useless. Although the gun is awkward to reassemble at this point—it's not easy to align a dozen parts all at once—it can still be done and the gun will still function. Proceed with the following steps, and it's "So long, NERF gun."

But in the interest of science, we pressed on. Using a hacksaw, we cut off the end of the white manifold to see what was inside. The lower part of the white manifold is a trough that leads to three ports, one for each barrel. Each has a small valve that protrudes into the trough. The trigger assembly slides into the trough from the other end. The assembly has an O-ring at its end to prevent air in the reservoirs from escaping. Cutting one of the reservoirs open, we found that the hose connects to a plastic tube inside.

Build Your Own

Here's a way to build a similar type of gun. This model is safe, provided you use it only to launch paper rockets—and provided you don't shoot them at people, animals, or breakable stuff. If your heart is set on being a menace to society, read no further.

The parts you need include a length of PVC schedule 200 1-inch diameter, a ball valve (not threaded), and an end cap that fits the pipe. Combined, these parts should cost less than six bucks. You will want to use PVC cement to secure them together. On your way home from the hardware store, stop at a bike shop. Ask for a used bike inner tube; stores usually have several they haven't yet thrown away. The valve stem is the part you need.

Drill a ⁵⁄₁₆-inch hole in the end cap and file it out just a bit so that the valve stem fits snugly through the hole. Warm up a trusty hot glue gun while you cut the valve stem out of the inner tube, keeping just a quarter of an inch or so of inner tube left around the base of the stem. Glob some hot glue around the base of the stem and jam it through the hole in the end cap. Hold the stem and end cap securely while the glue dries.

Cut two 1-foot-long pieces of the pipe and clean off the ends with sandpaper. Smear some PVC cement inside the mouth of the end cap and twist it onto the end of one pipe. At this point you will probably be getting a sense of the adequacy of the ventilation in your workspace; if it's not up to snuff, get some fresh air moving. Repeat the cementing treatment

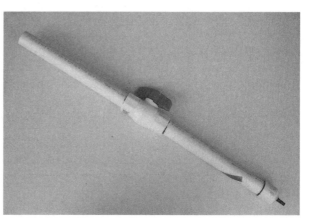

for each end of the ball valve, sticking one section of PVC pipe in each end. Let it dry for a few minutes and you'll be ready to test it.

Make a non-NERF rocket out of a sheet of paper. Roll it around the 1-inch pipe. Let the cylinder of paper loosen just a bit, then tape it with masking tape and remove it. Voilà: you've just made the fuselage. Fold over one end into a nose cone. Make sure it's airtight by blowing into the other end; if the tube holds the air, you're good to go. For longer flights, upgrade the nose cone by wrapping paper or card stock into a cone and taping over the airtight nose. Next, cut a business card in half diagonally to create two fins. Attach these at the base of the rocket.

To fire your gun, shut the ball valve, hook up a tire pump, and pump in 40 PSI (pounds per square inch). Slide the rocket onto the open end of the launcher and, taking careful aim to avoid hitting important stuff, twist open the valve to launch.

Science Experiments

Experiment with your invention by comparing the launch distances achieved at various levels of pressure or with fins of varying surface areas. You can also experiment to see how increasing the mass of the rocket (by adding paper clips) affects the launch distance.

NINTENDO
ENTERTAINMENT SYSTEM

History of the Nintendo Entertainment System

Nintendo got its start in the game industry in the late 19th century, by making playing cards. In 1975 it jumped into electronics by purchasing the rights to sell the world's first video game console, the Magnavox Odyssey, in Japan. Nintendo's engineers then started designing games for electronic arcade machines, many of which were great successes. (One of its most popular games was Donkey Kong.) Its next big move, in 1985, was to create a home video game platform, the Nintendo Entertainment System. Super Mario Brothers and Duck Hunt were two of the first games released for the NES. Nintendo sold some 60 million NES units and launched a revolution in the video game industry.

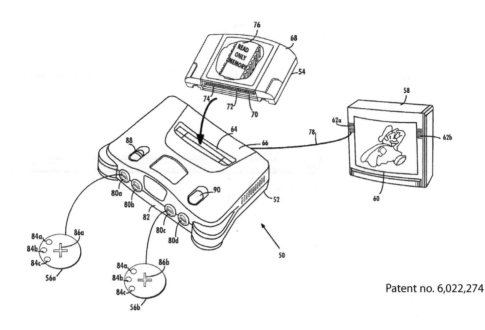

Patent no. 6,022,274

How Nintendo Entertainment Systems Work

Home video game systems consist of four main parts, one of which is your television. You get the other three when you buy the system: a controller, the console, and the game cartridge.

The console is a small computer with ports to connect cables to the TV and the controller. It includes a central processing unit (CPU) and memory. Memory is of two kinds: ROM (read-only memory) with instructions for the CPU, and RAM (random access memory) that records when you gain or lose points. The game cartridge is itself a ROM; it has the game programmed into its integrated circuits. Some cartridges also have RAM (random access memory) that allows you to record your scores from one game until you play again. The controller is a set of switches that send signals to the game computer. The video display resolution of NES is 240 by 256 pixels.

Inside the Nintendo Entertainment System

With all the newer machines on the market, an NES is easy to find, so you shouldn't have to spend big bucks to purchase one only to take it apart. They pile up at thrift stores, which is where we got ours.

We began our disassembly by taking apart a game cartridge. Nintendo's game cartridges are very difficult to open. Two clips in the front hold the two halves together, but the problem comes from three metal rivets. We wedged a large flat head screwdriver in to break them one by one. We know, we said not to use force in taking stuff apart, but in this case, brute force seems to be the only option.

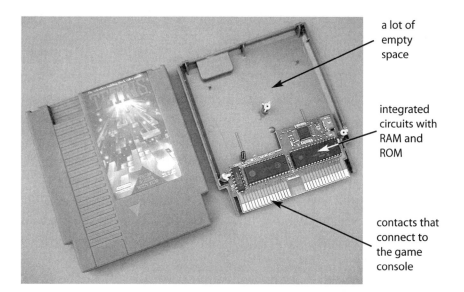

a lot of empty space

integrated circuits with RAM and ROM

contacts that connect to the game console

Inside, there isn't much. We looked for Mario, but he wasn't there. Instead we found four integrated circuits and a few other components. That's all. That's what you pay $50 for. Connecting the cartridge to the game console are 31 leads, or metal fingers. They fit into the port in the back of the console.

Remove four screws and the console opens. The electronics components are pretty uninteresting. What's neat is the tray that holds the game cartridge. Push it down—it stays down. Push it down again and it pops up to accept a cartridge. How does it know when to stay down and when to pop up?

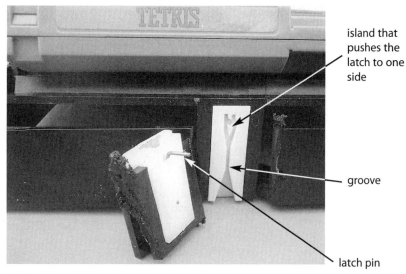

island that pushes the latch to one side

groove

latch pin

Latch

Of course, it doesn't *know* anything. But it's a neat mechanism. A metal latch (wire) sticks out, pushed by a spring, from the case. The latch rides in a groove mounted in the tray, which moves up and down. When the tray is up, the latch sits in the bottom of the groove. Pushing the tray down moves the groove past the latch. The groove is cut deeper on the right side, so the latch slides to the right. The groove is also cut deeper at the top, so once the latch has slid into the deeper part, it can't move back down the right side. It's caught in the center. The tray stays closed.

When it's time to change games, you depress the tray. The latch follows the contours of the groove to the left. As you release the tray the latch slides up a ramp on the left side until it's back where it started.

After recovering from all the excitement of the latch system, we checked out the power button. It stays in when you depress it once, and it pops out when you depress it again. But we'll leave it to you to take it apart.

Build Your Own

There are two different approaches to building your own gaming system. We'll call them the somewhat easy way and the extremely difficult way. The extremely difficult way involves finding microcontrollers and input devices (including joysticks, controllers, and video light guns). Then you must program the microcontrollers to interpret the signals from the input devices. After that, you can start thinking about what your game will do, how to play it, and how to program it into the microcontroller.

Most people who are interested in a homemade gaming system just want it to play games that they write; they don't want to mess around with the hardware. So the easier way is to turn to one of several companies that have done the hard work for you. You can find a few different programmable gaming systems at the XGameStation Web site (www.xgamestation.com). Once you've selected the system you want, it's just a matter of coming up with a great new game and programming it into your system. Even if you've never programmed, you can do it—there are a lot of resources available to help you learn how. The new games you create will not have all the flashy graphics you will find on a Nintendo Wii or PlayStation3, but the most popular games focus on game play instead of on graphics—remember Pac-Man and Tetris? Good luck creating your brilliant new game.

OPERATION

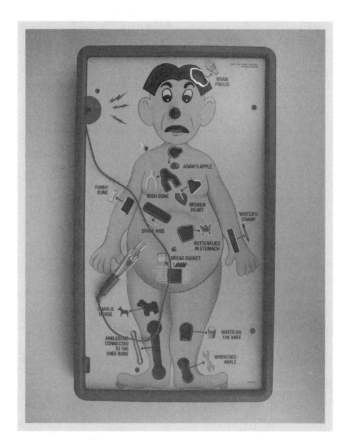

History of Operation

What started out as an industrial design class project became a successful game. Student John Spinello invented a game he called Death Valley in 1962. The game board had a desert motif, and the object was to poke a metal contact into openings in the board without hitting the sides of the openings. If the metal contact hit the metal sides, a loud bell sounded. Spinello sold the game to Milton Bradley in 1965, whereupon it went through a major transformation, but the core concept persisted.

Milton Bradley changed the "desert" to a human body and the metal contact to tweezers. The objective is to use the tweezers to remove identified maladies (in the form of small plastic trinkets) from openings in the body. Each player draws a card that indicates which

Patent no. 3,333,846

"operation" he or she has to perform and how much money the task is worth. Maladies requiring removal include: the funny bone, water on the knee, butterflies in the stomach, spare ribs, a charley horse, and more.

How Operation Works

The game is powered by two AA batteries. The negative side of the battery case is connected to the tweezers used to extract body parts. The other side of the battery connects the motor/vibrator, a light in Cavity Sam's (the patient's) nose, and the metal contact plate. If the tweezers touch the plate (exposed from the non-conducting game board) when a player tries to insert the tweezers into a body cavity, the circuit is completed, resulting in the motor vibrating the board and making a buzzing sound, and Cavity Sam's nose lighting up. This is a simple direct current circuit.

The mechanism behind Operation is similar to those of the old quiz games in which players tried to use electronic equipment to perform some memory-related task, such as matching states with their capitals. In that case, making the correct response meant putting one probe on an entry from one list and the other probe on the corresponding entry from the second list. Then the light or buzzer indicated whether or not you were correct.

In Operation, contact between the tweezers and the conducting plate completes the electrical circuit. Once the electrons flow, they power the light and the buzzer or vibrator. The vibrator is a small electric motor that spins an off-center weight. Like a tire that isn't balanced properly, the off-center weight spins around and causes everything attached to the motor to vibrate.

Inside Operation

Flipping the board upside down reveals the battery case and noisemaker. The same plastic device also holds and connects the light that shines as Cavity Sam's nose. It fits tightly into the molded holder on the bottom of the board. It will come out from the underside, but is much easier to take out when approached from the top.

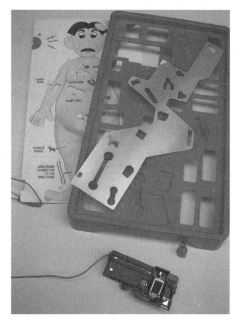

From the underside, you can see the shiny metal spring contact that carries electric current to the metal contact board. The other end of the circuit is a red wire that leads to the tweezers.

Turn the board upright so that Cavity Sam is facing up. With the blade of a flat head screwdriver, break the six plastic rivets. Done carefully, this won't destroy the game. Insert the blade between the cardboard and the plastic rivets, one rivet at a time. Leverage each rivet up so that it breaks off. The cardboard face of the game board (with the illustration of Cavity Sam) will come off to reveal the metal conducting plate underneath. This, too, will lift out. Then you can remove the black case that holds the batteries.

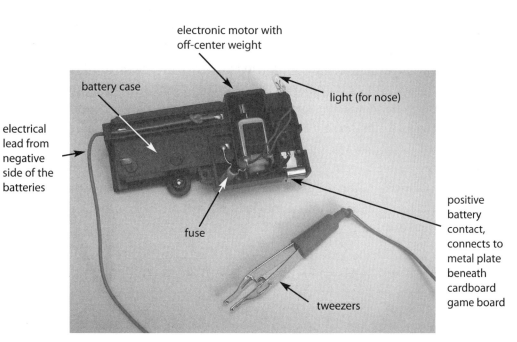

electronic motor with off-center weight

battery case

light (for nose)

electrical lead from negative side of the batteries

fuse

positive battery contact, connects to metal plate beneath cardboard game board

tweezers

Three small Phillips screws hold the cover on the case. By removing them, you can see the circuits, an electric motor, and the tiny lightbulb that lights up Cavity Sam's nose. The light is connected in parallel to the motor. A second lead connects one of the motor terminals to the positive side of the battery case. There is an electronic component between the motor and battery. The other motor terminal connects to the shiny metal spring tab.

The electronic component is a reset fuse. If the electric current gets too strong, the temperature inside the fuse rises, causing an increase in the resistance and limiting the current. As soon as the circuit current returns to normal, the resistance provided by the fuse drops to near zero and the game is back on.

Check out the motor. On the shaft is an eccentric weight, similar to the ones used as vibrators in cell phones. When the motor spins, it vibrates wildly.

Build Your Own

Start with a cardboard box. A shoe box works well. Cut the holes you want in the lid. Cover the outside of the lid with aluminum foil. Find the center of each hole from the aluminum foil side, and use a sharp pencil point to break the foil there. Fold the foil into the hole. You may need to tape or glue the foil edges to the underside of the lid.

Purchase AA batteries, as well as a two-battery AA cell battery holder, some clip leads, and a buzzer or a 1.5-volt lightbulb and socket (available at Radio Shack and other electronics stores). Connect one clip lead to the aluminum foil and to one of the battery holder terminals. Connect the other clip lead to the other terminal and to one side of the buzzer or light socket. A third lead goes from the other side of the light or buzzer to a metal probe, such as a nail or metal tweezers, that you'll use in your game.

Put a piece of card stock on top of the aluminum foil, decorate it appropriately, tape it in place, and punch out the holes, leaving foil exposed along the edges of the holes. You could make a gold mining game (extract the nuggets), a diamond mining game, or something really different. To hold the items you're trying to extract, cut the bottom of the shoe box down evenly on all sides so it is about 2 inches tall. With the lid in place, the objects will be about 1 inch below the top surface.

For the retrievable objects, you may want to use nonconductive items similar to Operation's plastic trinkets. Using materials that conduct electricity may make the game a bit more challenging, because you will have to keep not just the metal probe but the objects themselves from contacting the foil.

PLAY-DOH

History of Play-Doh

Play-Doh originated, believe it or not, as a product used to clean wallpaper. The genius in Play-Doh was not in its invention, but in recognizing that the cleaner could also be used as modeling clay for kids.

Years ago, people used a flour-based dough to clean the coal-fired soot off their wallpaper. In 1955 Joseph McVicker sent a sample of it to his sister, a classroom teacher who was looking for a better modeling product. The wall cleaner was an improvement over what she had been using as modeling clay, and soon other schools also started to use it as well. McVicker formed a company, Rainbow Crafts, to make and sell the stuff, and he soon became a millionaire. The company later merged with Kenner, which eventually was sold to Hasbro.

Patent number 3,167,440 was awarded to Joseph McVicker and his brother Noah in 1965, for the "plastic modeling composition" that became known as Play-Doh. The patent specifies some 20 different recipes for making modeling composition and the process the inventors recommend, but it doesn't identify which, if any, of these recipes is the one that's actually used to make Play-Doh.

People around the world have purchased many hundreds of millions of cans of Play-Doh, and its distinctive scent is one of the most recognized smells among people of all ages in the United States. Nearly 100 million cans of Play-Doh are sold each year.

How Play-Doh Works

Play-Doh is vegetable flour with a bunch of other stuff mixed in. It is a mixture of water, wheat starches, salt, and several other chemicals that give it color and its unique smell, and that help keep it moist. A petroleum additive gives the dough a smoother feel. Borax prevents mold from colonizing in the dough. Play-Doh must be sealed in an airtight canister in order to prevent its water content from evaporating; leave it out of its can overnight and you'll end up with a hardened figure.

Inside Play-Doh

It's dark. Very, very dark.

Make Your Own

Heat a mixture of 1 cup flour, ½ cup salt, 2 tablespoons vegetable oil, 1 cup water, and 2 tablespoons cream of tartar on the stove at low heat, stirring constantly until the mixture becomes a sticky mess. Transfer the mixture to a plate. When it has cooled enough to touch, knead it. If you want to, you can knead in some food coloring. It's now ready for your creative ideas. When you're not playing with it, store it in an airtight container or plastic bag.

Following this recipe may raise the question "What is cream of tartar?" It's potassium bitartrate, $KHC_4H_4O_6$. It's used in recipes and in making baking powder. It's also a laxative . . . so don't eat your Play-Doh.

Playful Penguin Race

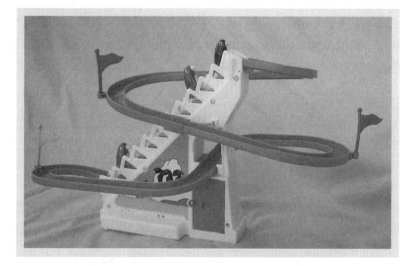

History of the Playful Penguin Race

This iconic toy was created by the Dah Yang Toy Company of Taiwan. Mr. Chie-Teh Yang founded the company in 1958.

How the Playful Penguin Race Works

One motor, powered by a D cell battery, operates the toy. Plastic penguins ride an escalator to the top, then slide down a twisting ramp, only to start all over again. (Perhaps a metaphor for the sometimes fruitless struggle to climb the ladder of success.) One of the great mysteries of this toy is the penguins' escalator trip up. Rather than stand on the steps, the penguins rest on pins (at wing level) that are held up by the two sets of plastic rails. One set, made of clear plastic (making it harder to see), raises them up to the next level. When penguins reach the top of a step they lurch forward off the clear rails onto the outer, white rails and slide forward. The clear plastic rails move down and out of the way. While the penguins slide forward, the next step in the clear rails oscillates down to pick up the bird to carry it to the next step.

Watch how the penguins lean backward at the start of each cycle, then lean forward as they rise. They stub their feet on the steps as they rise, and this causes them to lean forward. When they clear a step, their tail fins rotate forward and their heads tip back. At the top of the escalator, the penguins roll onto the ramp on their way back to the start.

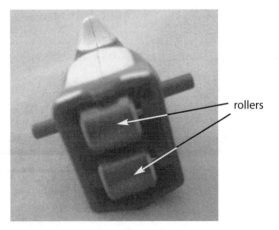

rollers

Check out the bottoms of the penguins. They roll on two wide rollers that make them stable.

Penguin "feet"

The obnoxious sounds you hear as you operate the toy are generated by a reed horn in the base. The motor that lifts the rails to push the penguins up the escalator also punches a bellows to force air through the reed horn.

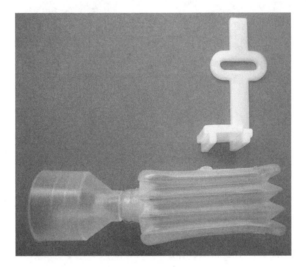

Inside the Playful Penguin Race

To take this toy apart, first remove the tracks. There are three sections. A few screws hold the two halves together; once you remove them, you can lift off the half that holds the battery.

The motor inside drives a cam (the wheel with an off-center peg) that pushes the clear plastic rails in a circular motion. You can lift the rail assembly off the two pins that hold it.

The squawker is below the motor. Pull it out to see how it makes sounds. A white plastic arm pushes on the plastic squawker to force air out. When the arm lets up, air rushes back into the bellows. Air entering the bellows passes through a reed that makes the noise.

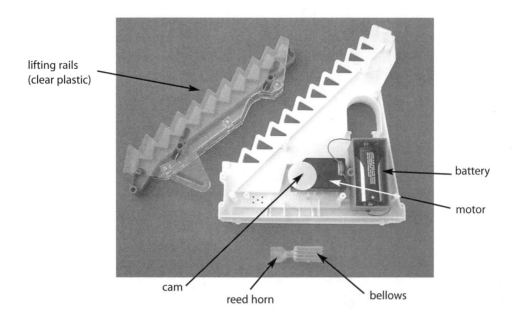

And what moves the plastic arm? Pull the motor out to find a second white plastic cam. It is attached to a different gear inside the motor/gearbox, and it rotates at a faster pace than does the escalator. The result is more squawks than moves. (Another metaphor for life?)

Build Your Own

Recreating the "up" part, which would require motors, is too complex to discuss here. Making the "down" part is easy. Kids and those who are young at heart enjoy rolling balls down raceways or making their own raceways to roll balls down. PVC pipes, cardboard mailing tubes, and rolled-up newspaper all make for good raceways. Tape a raceway to a large cardboard box, the refrigerator, or a 4-foot by 8-foot sheet of Peg-Board or press-board (they're both cheap and provide a good surface to attach stuff to). See how elaborate a raceway your design team can make.

Resources

Bernie Zubrowski wrote the ultimate book on building raceways: *Raceways: Having Fun with Balls and Tracks* (William Morrow and Company, 1985).

Pop-Pop Boat

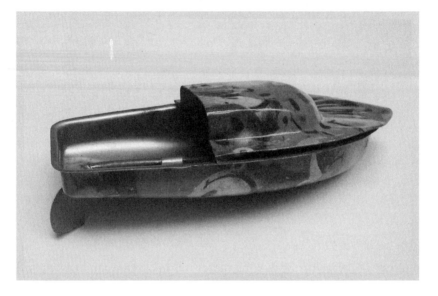

History of the Pop-Pop Boat

Thomas Piot, a French electrician and engineer living in London, obtained a British patent in 1891 for a "water pulse" engine, which is the basis for steam-powered boats with no moving parts called pop-pop boats. In Piot's design, water is heated in a coil of copper tubing. Charles McHugh, an Indianapolis inventor who later invented a device for roasting grain, improved the design in 1916 when he substituted a diaphragm boiler for the coil. The patent shown here is McHugh's (patent number 1,200,960). These toy boats are becoming more popular today.

How Pop-Pop Boats Work

To operate the boat, you fill the boiler with water. Usually the boat comes with a water dropper, and you use it to squirt water into one tube until it fills both sides. The only reason there are two tubes (instead of one) is so you can easily fill the boiler with water.

Now, place the boat in water and add heat, either from a tiny candle that comes with the boat or from olive oil and a wick. After about a minute of heating, the water inside the boiler reaches the boiling point. As water turns to steam—it occurs so fast that the device is called a flash boiler—the steam takes up a much greater volume (it is now a gas

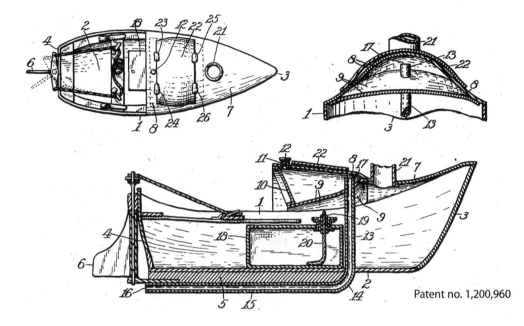

Patent no. 1,200,960

rather than a liquid) and pressure builds quickly. The high pressure forces water out the two tubes, propelling the boat forward. A candle or a half-teaspoon of olive oil will keep it going for 30 minutes.

Of course, if that were the entire story the boat would move a few inches and stop; the boiler would be out of water and the steam engine would quit. But then something neat happens. Once the boiler has released the pressure, cooled, and reverted to a low-pressure state, it sucks water into the two tubes, which are underwater. With the candle or oil still heating it, the small amount of water in the boiler boils in a flash, forcing steam and water out of the tubes. Each release of pressure creates forward motion and the "putt-putt" sound.

The boat works with either the more basic coil of copper tubing or with the boiler you see in boats you purchase. The boiler has a flexible top (called a diaphragm) that is made of a thin piece of copper. The diaphragm flexes up with high pressure and springs down with low pressure. You can hear the click as the diaphragm flexes on each cycle. This action helps force water and steam out and helps draw water in during the intake part of the cycle.

Why does the boat go forward? After all, it pushes water out and sucks water in at the same place—don't the forces balance out, with no net motion achieved? Obviously not, but why? Apparently water is propelled out the tubes, pushing the boat in the opposite direction, but during the suction side of the cycle, water is drawn from all around the tube—not just immediately behind it. Thus much of the reactive force generated during the "in" cycle balances itself out and doesn't pull the boat backward. To test this phenomenon, try the experiment of first blowing out a candle, then trying to "suck out" a candle

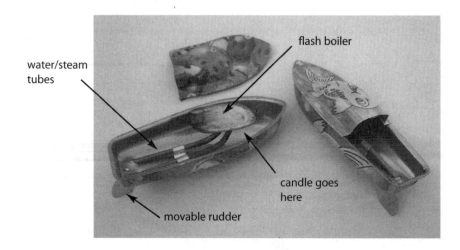

water/steam
tubes

flash boiler

candle goes
here

movable rudder

by drawing breath. The former is easy to do, the latter all but impossible. (If you intend
to try this experiment, keep some Vaseline on hand just in case your lips get too close to
the flame. Explain *that* injury at work!)

These boats come equipped with a movable rudder. Push it to one side and the boat
will turn in circles. As the boat moves forward, water hitting a rudder turned toward the
right pushes the stern of the boat to the left. When the stern moves to the left, the bow
moves to the right as the boat pivots around its center of resistance.

Inside the Pop-Pop Boat

What's truly amazing is that a toy with no moving parts can propel itself so well. Remove
the deck by prying carefully with a flat head screwdriver. The edge of the deck is folded
around the lip of the hull, so you want to loosen the edge. With just a little effort, the deck
will fly off. Be careful where you point the screwdriver while prying, because it will likely
continue in that direction if you slip.

Inside the boat you'll find the boiler and two tubes leading to it. That's all. We don't
recommend opening the boiler—there isn't anything to see, and you will have difficulty
getting it back on and sealed properly. Below the boiler there is, of course, a heat source:
either a candle or reservoir for oil.

Build Your Own

We'll focus on the steam engine only; you can figure out how to build a boat hull. Create
a coil from 12 inches of ³⁄₃₂-inch (outer-diameter measurement) copper or aluminum tubing. We purchased the tubing at a hobby store. Carefully start to bend the tubing around

a ½-inch dowel. Once you start doing this, put the tubing beneath the dowel. Have someone push down hard on the dowel while you bend the tubing up around it. Coil the tubing two and a half times around the dowel.

Position the coil inside the hull so you can get your source of heat underneath it and so the two tubes protrude through the bottom of the hull. The propulsion of our model wasn't as strong as that of the diaphragm model we bought.

Resources

The Flying Circus of Physics by Jearl Walker (John Wiley, 1977) describes how pop-pop boats work, and it is a great general reference on science.

Potato Gun

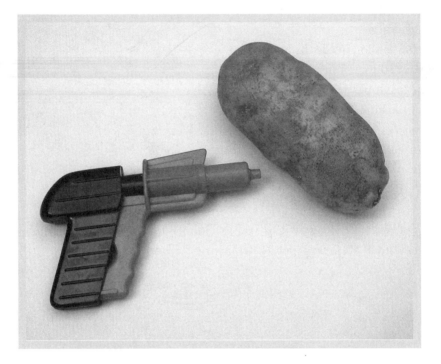

How Potato Guns Work

Aside from the chance encounter with someone's eyeball, this is a harmless, albeit messy, device that launches nibs of potato through the air. To operate this technical marvel, jab the barrel into a spud. Twist off a nib of potato and withdraw the barrel. Take aim at some unsuspecting inanimate target and squeeze the trigger.

The nib wedged into the barrel blocks air from leaving the gun. As you squeeze the trigger, you bring the two plastic halves of the gun together, driving the piston farther into the barrel. The piston seals the narrow end of the barrel shut, preventing air from escaping out the back, while the nib prevents it from escaping out the front. As the piston continues forward it pushes this relatively large volume of air into the much smaller cylinder that holds the nib, increasing the air pressure. As the pressure builds—this all occurs in a

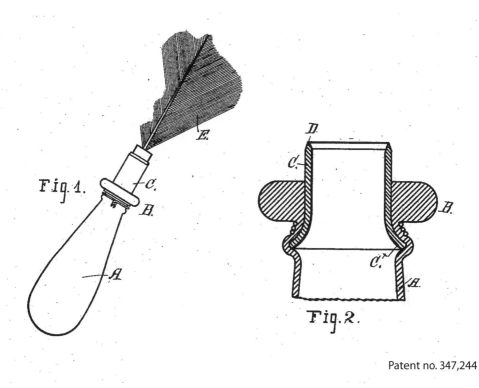

Fig. 1.

Fig. 2.

Patent no. 347,244

fraction of a second—it eventually exceeds the frictional forces that hold the nib in the barrel. Out comes the nib, on a flight of 20 to 30 feet.

After the gun is fired, a spring in the wider part of the barrel pushes the two halves of the gun apart, withdrawing the piston from the cylinder. You're ready to shoot again.

We took a good look at our potato gun, and it seemed to us that the barrel was too short—that if were lengthened, the nib would stay in contact with the compressed air longer and would fly farther. So we add a section of fat straw to the end of the barrel. We even tried a couple different lengths of straw. But it shot no farther.

Inside the Potato Gun

This is easy. Grasp the handle (black in our model) with one hand and the barrel (ours is red) with the other and pull them apart. (They go back together just as easily.) The only part of the gun that you could misplace is the spring that fits around the piston. In addition to the handle, the barrel, and the piston, there is also a white gasket that sits on the end of the piston. Four parts total—not counting the greasy substance dolloped onto the end of the piston (be careful where you put the parts down).

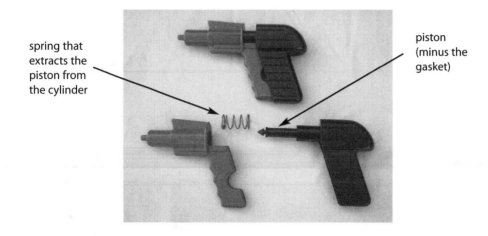

spring that extracts the piston from the cylinder

piston (minus the gasket)

Build Your Own

You can make a simple spud launcher from metal pipe that is used for electrical conduit. The pipe is called EMT, and it is sold at most hardware stores. Buy a piece of ¾-inch-diameter EMT and the longest ½-inch-diameter wood dowel that will fit inside the pipe. Using a hacksaw, cut the pipe so that it's an inch shorter than the dowel.

With the EMT standing vertically against the ground, hold a spud over the upper end. Pound the spud down, onto and into the EMT. When the pipe sticks through the potato, turn the EMT over and repeat the process so that you have a plug of spud in each end. One spud is the projectile; the other is the wadding for the piston.

Jam the dowel, which acts as the piston, into one end of the EMT. Once you have it started into the barrel, turn the launcher upside down so that the dowel rests on the ground. With a mighty push, pull the EMT down to the ground around the piston to send the spud across your yard. A good launch will go 50 feet.

There are many designs for spud guns—gizmos that will launch an entire potato. We've built several, both combustion and pneumatic. The simplest pneumatic launcher uses a PVC ball valve glued between a barrel and a chamber. The chamber is sealed with a PVC end cap. We glued a bike tire valve stem into the inside of the end cap before gluing the cap onto the chamber tube. Closing the valve lets you build up pressure with a bike pump. Opening the valve releases the air pressure to launch the spud. (See our instructions for building your own NERF-type pressure gun, page 85, for a more detailed explanation of the valve component.)The Internet has many sites devoted to design and operation of much more complex launchers, including the Spudgun Technology Center (www.spudtech.com) and AdvancedSpuds (www.advancedspuds.com).

Combustion spud guns use barbecue sparkers to ignite combustible fluids, usually hair spray, in a chamber. The explosion, which can send a flame out of the barrel—very impressive at night—can easily hurl a potato 100 yards.

Pullback Car

History of the Pullback Car

Toy cars and trains powered by the energy stored in a spring or rubber band have been common for well over 100 years. The first toys were spring windups. Later, rubber bands were wound by twisting a wheel or propeller.

The first pullback toy vehicle, a wagon, was patented in 1878 by F. R. Hoard (patent number 202,651). Pushing or pulling the wagon stretched a spring that was attached to the large rear wheels. On release the spring would contract, driving the wagon forward.

The invention of the disengagement gear made the modern pullback car possible. Japanese Inventor Akio Masuda got a patent (number 4,116,084) in 1978 for this device, which continues to be used in many different toy vehicles today.

How Pullback Cars Work

When you pull the car backward you store energy in its internal spring. When the car is released, the energy in the spring drives the wheels. What's confounding is that you only need to pull the car back a short distance to get it to travel so far forward.

The trick here is that there are two different sets of gears. Traveling forward, the spring turns a large gear that meshes with a smaller gear. This big gear/little gear combination is repeated twice more. As a result, one rotation of the shaft attached to the spring yields several rotations of the wheels. The unwinding spring will drive the car at a fast rate of speed.

When the user pulls the car back, a different set of gears lines up that requires only a few backward rotations of the wheels to fully wind the spring. But how does the car switch gears? One of the gears has room to move, and with backward rotation, it slides out of the way so other gears wind up the spring. When you release the car, the gear slides back into position and engages.

Inside the Pullback Car

Remove the one or two screws beneath the car to separate the body from the chassis. Some cars have plastic pieces molded to look like seats, steering wheels, etc. These fall out when you open the car. Underneath the interior are the two axles.

The rear axle is connected to the drive gears. Before opening up (and probably destroying) the gear assembly, watch it while you drag the wheels backward across the floor. The gears quickly wind up the spring (inside the cylindrical plastic case). When the spring is wound, the gears disengage to prevent you from overwinding and breaking the spring. The clicking sound you hear after you've pulled the car back a few inches is two gears sliding past each other.

spring encasement

slide that disengages

Rear wheel assembly

Also notice that the gears change when you pull the car backward. Along one side of the gear housing you can see an oblong opening where the axle that is attached to a gear sits. This opening allows the gear to move up and down. The other end of the gear shaft is secured in a hole just large enough to admit the shaft. The gear kicks out of the drive train when you pull the car backward, allowing a different set of gears to wind the spring. This set of gears has a big gear driving a little gear (connected to the spring) to wind the spring with a few turns of the wheels. When you release the wheels, allowing them to spin, the gear that had moved out of the way now moves back into position. The forward gear train has a large wheel on the spring shaft driving a smaller wheel. This combination produces fast speeds and a long traveling distance.

To explore deeper, take one rear wheel off the axle. Remove the wheel that is on the opposite side of the spring. You may need pliers to grab the wheel and pull it free of the axle. The wheels are held on either by pressure or with some glue.

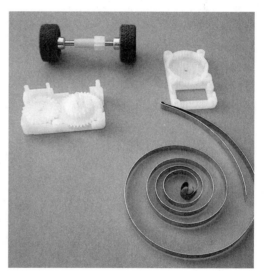

Using a small screwdriver, you can pry under the plastic piece (gear housing) that is on the same side as the wheel you just removed. Break the three plastic rivets that hold it on; be prepared for gears to fly across the room. Where did you put those safety goggles?

If you opened it without launching the gears, remember where each gear goes so you can reassemble the gears to see them operate. Remove the remaining gears. Behind the last gear is the spring. If you pull this out it will unwind, and it may be difficult to get back into its housing.

Build Your Own

The closest we've come to replicating this spring-driven car is a rubber band car or a mousetrap car. Both put a lot of stress on whatever is holding the drive axle in place. You can drill holes through a piece of wood (the car chassis) or screw in screw eyes. The non-drive axle can be secured by gluing a straw to the bottom of the chassis. Run a dowel through the straw and attach wheels. The rear of the chassis needs to be cut to allow a rubber band or string from a mousetrap to wrap around the axle. Drill a hole in the top of the car to mount a short piece of dowel to hold the other end of the rubber band.

For a mouse trap car, glue an old-fashioned mousetrap to the top of the chassis. Remove the catch and the release bar that keep the trap set until an unlucky mouse wanders by. Cut off the bail—the metal wire that whacks the mouse—so you can slide on a piece of brass tube (available at hobby stores). The tube is used to extend the reach of the trap. Glue fat straws on the bottom of the chassis and run dowels, as axles, through the straws. For wheels, attach some of those CDs that seem to arrive, unsolicited, every week in the mail. Inserts to connect the CDs to the axles are available from the KELVIN and Doc Fizzix Web sites (see Resources). Tie one end of a string to the top of the brass tube, and wind the other end around the rear axle while holding the trap open.

Simple, right? In truth this project may sound easier than it is. There is a real art to making a great mousetrap car. For more information, see Resources, below.

Resources

Wheels and axles, along with mousetrap kits, can be purchased from the KELVIN Web site (www.kelvin.com) and Doc Fizzix (www.docfizzix.com).

Doc Fizzix's book is a mouse trap car design engineer's dream: *Mouse Trap Cars: A Teacher's Guide* by Al Balmer and Mike Harnisch (Doc Fizzix Publishing Company, 1998).

PUSH 'N' GO CAR

History of the Push 'n' Go Car

Sadly, we don't know the origins of this wonder. Whatever its story, one thing is certain: it has been a very popular toy for many, many years.

How Push 'n' Go Cars Work

You push; it goes. By depressing the driver's head, a spring is compressed, storing energy to power the car. When the car is released, the spring pushes the driver upward. Potential energy stored in the spring becomes kinetic energy that turns the wheels.

The driver's base has ridges that engage the first of four pairs of gears that drive the wheels. The gears are paired so that large gears drive small gears—this converts the modest vertical motion of the driver into the car's substantial horizontal motion. Even if the rolling that occurs after the spring has fully extended is disregarded, the car moves much farther forward than can seemingly be accounted for by the one inch of motion you put into it.

The mechanism behind these cars is so interesting that we use them in museum exhibits and as workshop challenges for graduate students in engineering and physics. It's an engineering challenge to figure out how the Push 'n' Go works before taking it apart. You can estimate the gear ratios by measuring how far the car travels while powered by the spring and comparing that distance to the distance you depress the driver. To get in even the right ballpark, you need to know that the gears are driving a wheel that has a diameter of about 1⅝ inch, or a circumference of 5.1 inches. Each rotation of the axle, therefore, pushes the car 5.1 inches.

Inside the Push 'n' Go Car

This family of toys is ideal for taking apart. A few screws hold a car together, and they go back in, so reassembly is easy—provided you don't open the gear assembly.

Play with the car before taking it apart. You push down the driver's head about 1 inch to make the car go 10 feet or farther. How does a 1-inch push result in such a great distance traveled? When you push down, the rear wheels don't move. However, when you release your hold on the driver's head, the rear wheels spin. What sort of mechanism turns the wheels in one direction, but not the other? What stores the energy you input (by depressing the driver's head)? Taking apart the car will reveal the answers.

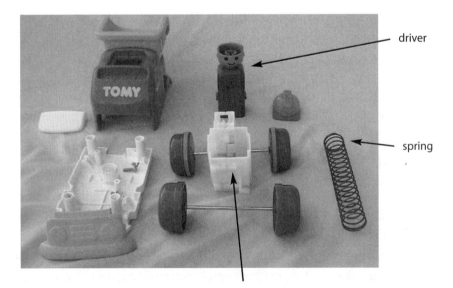

driver

spring

gear encasement

Four screws, accessed from beneath the car, hold the body to the chassis. When they come out, lift off the body. The energy storage device for this toy, a spring, lifts out, as does one of the axles—depending on which Push 'n' Go model you have. In some models you need to remove the driver's hat to remove the driver and the pedestal he rides on. The spring resides inside the pedestal.

Take the gear assembly off the chassis. If you decide to go beyond this point in your disassembly, take care to remember where components go. The drive wheels have rubber tires (to allow for better traction on slick floors) sandwiched between plastic parts. By removing the two screws on the insides of the wheels, you can take the wheels apart. This takes some prying underneath the tire to separate the wheel parts.

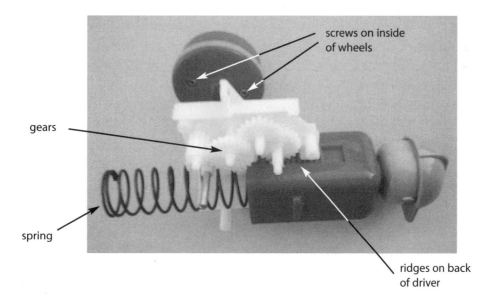

screws on inside of wheels

gears

spring

ridges on back of driver

Note the ridges on the back of the driver's pedestal. They mesh with a gear that pro-trudes into the spring chamber. Below the ridges there is an open space that lets the wheels continue to turn even after the spring has extended fully. This allows the car to roll; if there were ridges all the way down, the car would be forced to stop when the pedestal was fully extended. So the reason the car can roll 10 feet is that it builds up momentum while the spring is expanding (and the driver's head is rising), then the gear disengages from the pedestal, allowing the car to keep rolling.

Slide a finger into the chamber (where the driver's pedestal goes) to spin the gear. The wheels don't spin when you push downward. When the first gear is pushed on, it lifts the second gear away from the gears that follow. On the sides of the gear assembly you can see that slots that hold the second gear aren't circular—they're elongated to let the gear move. This elongated slot allows the gear to disengage on the downward push and reengage when the spring is lengthening.

What would happen if the designers hadn't let the second gear disengage? The wheels would spin madly in reverse when you compressed the spring. When the spring was released, the wheels would reverse direction, placing considerable stress on the plastic gears.

Pull your finger up to spin the first gear. The second gear engages the third, and the wheels spin. A small spin of the first gear causes the wheels to move far. The car moves forward as the pedestal rises, rotating the gear—all forced by the compressed spring.

Four screws hold the gear assembly together. If you choose to open it, study the locations and orientations of the gears before proceeding. This is a toy you can fully reassemble. If you use care, you can take it apart and put it back together many times.

Radio-Controlled Car

History of the Radio-Controlled Car

One year after Guglielmo Marconi received the first patent for the radio, electricity pioneer Nikola Tesla was awarded his own patent (number 613,809) for radio control of a boat. Tesla demonstrated his radio-controlled boat at Madison Square Garden in 1898. His system used a switch on board the boat that he could control with radio signals.

Not much was done with R/C systems until World War I, when the German Navy created R/C boats to ram enemy warships. World War II saw an expansion of the uses for R/C devices, but nonmilitary uses for R/C were limited until solid state electronics were invented, starting with the transistor in 1947. One of the first commercial uses for the technology was remote-controlled garage doors. Later came R/C toy cars.

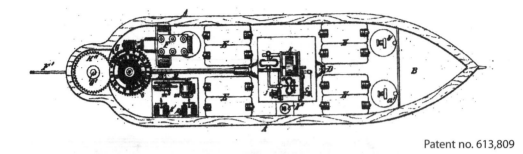

Patent no. 613,809

How Radio-Controlled Cars Work

This toy's two components, the handheld controller and the car, both require batteries. Usually a 9-volt transistor battery is used in the controller, and some combination of batteries is used in the car. The controller is a radio transmitter that sends pulse-modulated signals to the receiver inside the car. Pulse modulation means turning on and off a radio wave to create a pulse. A particular series of pulses makes up a command to go forward or reverse and to turn left or right. A receiver in the car picks up the pulses and sends them to an onboard integrated circuit (IC) that interprets the train of pulses and sends electrical current to either or both of the motors.

The Federal Communications Commission has set two radio bands for toy model cars: 27 MHz and 49 MHz. Radio-controlled toy planes operate at 72 MHz. Toy boats and other non-airborne models can operate at 27 or 75 MHz.

Pressing a button on the controller signals the integrated circuit inside it to send a series of pulses to the receiver. The first in the series is a wake-up call to alert the receiver that a directive is coming. This is called the synchronization signal, and it consists of four pulses of the same duration. Following the synchronization signal comes the command signal, which consists of some number of pulses. The integrated circuit in the car counts the number of pulses to determine what the command is. For example, 16 pulses might signal the car to move forward.

Because the inexpensive electric motors in radio-controlled cars run at speeds too high to power the wheels directly, the cars have a gear system to modify the speed. By counting the number of teeth on each gear you can calculate the gear ratios. For example, if the gear attached to the motor shaft has 10 teeth, and if it's connected to a gear with 20 teeth, the gear ratio is 1:2. That means that the axle supporting the larger gear turns half as fast as the motor turns. Typically there are several gears in the drive train to slow down the speed of rotation.

Inside the Radio-Controlled Car

The handheld controller doesn't have much of interest to see inside. The switches are the most interesting parts. R/C controllers have a variety of switches: steering wheels, toggles, and joysticks. The switches are connected by wires to the circuit board, which also connects to the battery. The controller has a transmit antenna.

Inside the car there's a lot to see. Most cars have a handful of screws that hold the body to the chassis. Removing them allows you to see all the cool stuff inside. Like the controller, the car has a circuit board and an antenna. Cars that are designed to flip over have their antennas wrapped inside the car body to keep them safely out of the way. Typically there are two motors: a larger motor to drive the rear wheels and a smaller motor to power the steering. However, there are lots of variations on this scheme.

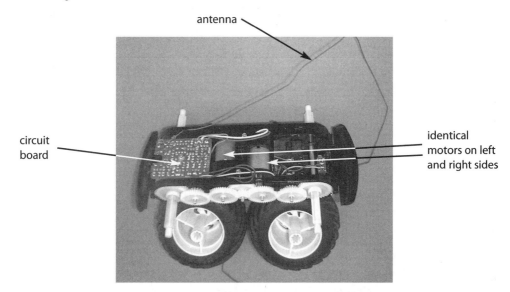

The Tyco car shown above has two identical motors. One powers both the front and rear wheels on one side and the other powers the front and rear wheels on the other side. The ability of one side's wheels to move in reverse while the other side's wheels are going forward allows the car to spin around in place. Gears are arranged along each side of the car to power front and rear wheels equally.

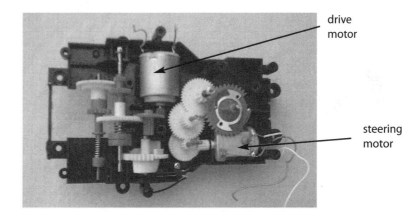

Other cars have a transmission that you can set for high or low speed. You slide a lever, and that pushes different gears into position to drive the car.

More expensive cars have functioning differentials. The differential, which is similar to the one in your family car, allows the two rear wheels to turn at different speeds. When rounding a corner, the inner wheel follows a smaller circle than does the outer wheel. The differential allows the inner wheel to turn more slowly. The motor drives the larger outer gear, which is called a "crown wheel." Shafts on either side of the differential convey power to the wheels. Inside, the smaller gears spin when the car turns; by spinning, they allow the two axles to turn at different speeds.

Build Your Own

To make your own radio-controlled car, purchase the R/C controllers and motors from a hobby store and add them to a model you make.

You can make more basic electric-powered cars using electric motors you purchase at electronics or hobby stores. Or you can extract the circuit board and the motors from an old R/C car. Label the wires going to the circuit board so you know which wires control the motors, which wires provide power, and which wire goes to the antenna. Try connecting new motors and seeing if the circuit board and hand controller work. If you purchase several cars and controllers at a thrift store or yard sale, or from another inexpensive source, you're likely to end up with enough working parts to control a model you make from scratch.

Resources

KELVIN (www.kelvin.com) is a good source for high-speed DC motors and gear-head motors. See Ed's book *Loco-Motion: Physics Models for the Classroom* (Zephyr Press, 2005) for simple design ideas for model cars.

REMOTE-CONTROLLED ROBOT

History of the Remote-Controlled Robot

We don't know the origins of this toy, but we suppose it was created long, long ago in a distant galaxy. Anyway, it has the Force—at least when batteries are inserted.

How Remote-Controlled Robots Work

This battery-powered toy, controlled remotely via a cable, rolls along the floor while rotating its dome and lighting an LED (light-emitting diode). The operator pushes a button that sends electrical power to the motor inside, and that drives the wheels through gears. The gears for the right and left sides are different, which allows the robot to change directions while backing up. By inserting additional gears into the train, one wheel receives forward rotation while the other receives backward rotation. The ability to have the robot change directions using only one motor saves the manufacturer the cost of a second motor.

Many people confuse *remote* control and *radio* control. An example of remote control in action is the mouse that is plugged into your computer. You are controlling the computer without touching the computer. If you extended the mouse cable you could control it from a more remote location. The wire connecting the control unit to the robot carries electric current to power the single motor inside.

Radio control involves sending and receiving signals encoded on radio waves. In this case, there is no physical contact between the control unit and the device being controlled. Radio-controlled toys carry the batteries that power their movement. In remote-controlled toys it is common for the batteries to be in the toy's handheld control unit.

Inside the Remote-Controlled Robot

The remote control comes apart to reveal how simple the controls are. A spring provides some resistance to the thumb knob. The knob rotates up and down to move a switch, which runs the robot forward and in reverse and also stops it.

left axle

speaker

right axle

The side panels of the robot itself come off the two legs; remove four screws on the backs of the legs to open them. Underneath the panels are the gears that drive the front wheels. The panels help hold the front and back halves of the body together.

Inside the body there is a speaker so that your R2D2 can emit squeaks and squawks. The circuit board next to the speaker generates the sounds. Inside the black plastic case are the gears that both drive the robot and turn its dome. What's especially neat is how the robot changes directions.

Many simple remote-controlled cars change directions by backing up. Typically the front wheels pivot when backing up to change orientation. But in this robot the gears shift.

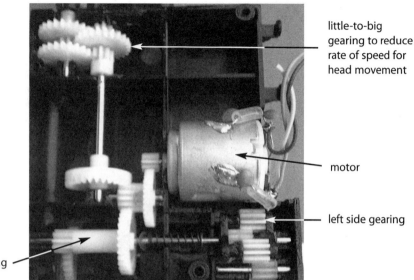

little-to-big gearing to reduce rate of speed for head movement

motor

left side gearing

right side gearing

The right drive wheel is linked directly to the motor and changes direction as the motor changes rotation. The left wheel is connected through a differential gearing so it always rotates in the same direction. The differential does this by changing the number of gears that are in line with the wheel. A wheel that's connected to just one gear rotates in the opposite direction than a wheel that's connected to two gears. So to keep the left wheel moving forward when the motor has reversed, the differential shifts between three pairs of gears and four as its shaft changes direction.

The same motor turns the dome or head. The pairings of gears, with the driving force going from little gears to big, slows down the rotation for the dome. The dome doesn't spin in circles; it rotates back and forth. A cam in the drive shaft changes the shaft's round-and-round rotation of the gears into back-and-forth rotation.

RUBIK'S CUBE

History of the Rubik's Cube

This colorful, confusing cube was invented in 1974 by Erno Rubik of Hungary. Originally called the Magic Cube, it was released in Budapest in 1977. Ideal Toy Company licensed it, renamed it after its inventor, and introduced it to the rest of the world.

What started off as a simple logic toy that focuses on spatial awareness quickly turned into a worldwide fad. Between 1980 and 1982, more than 100 million Rubik's Cubes were sold. The British Association of Toy Retailers was so impressed by the device that it named the Rubik's Cube Toy of the Year in both 1980 and 1981.

Over time, demand grew for more challenging Rubik's Cube puzzles, so Erno Rubik and others designed them. Aside from the traditional 3x3x3 cube, there is also a 2x2x2, a 4x4x4, and a 5x5x5. In 2005 a 6x6x6 was invented, followed a year later by a 7x7x7. That, however, may not be the last of the Rubik's Cube innovations; the 7x7x7 has already been solved in less than six and a half minutes. There are currently plans for making a cube with up to 11 blocks per row.

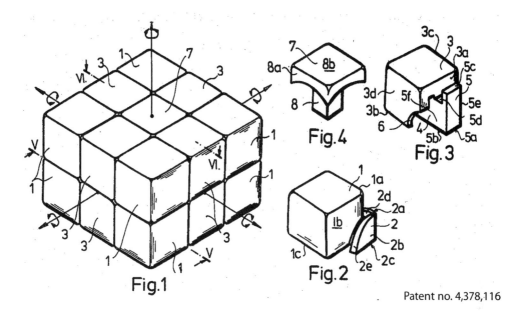

Fig.4

Fig.3

Fig.2

Fig.1

Patent no. 4,378,116

Inside the Rubik's Cube

Taking apart a Rubik's Cube may seem a bit of a challenge. You aren't presented with any of the usual screws, rivets, or joiners. Instead, the entire cube is snapped together. To take it apart, all you have to do is pry. The easiest way to start is to rotate one plane of the cube 45 degrees, then pull up on the corner piece. It takes a lot of force to pop out the corner piece, but be careful: too much force and you could break it beyond repair. Once the first piece is out, you can rotate the remaining planes and pull them out as well.

Although the 3x3x3 cube appears to be made of 26 smaller cubes (or 27, if you assume there's one more in the middle you can't see), there are only 21 pieces, and none of them are really cubes. There are eight corner pieces—which look like cubes with connector tabs on one corner—and 12 edge pieces—cubes with the connector tabs on their inside edges. The final piece, the core, is the most interesting. It comprises the center cubes of all six faces, which are connected by three intersecting axes that allow each cube to rotate separately. The other 20 pieces slot in around the six center cubes, and their tabs allow them to stay together as they rotate.

How It's Solved

There are a lot of possible positions in which to find your Rubik's Cube—*lots!* The 3x3x3 cube's packaging usually indicates that there are a billion possible positions. If that were true, and if a person could try one position per second, it would take that person about 33 years to get through them all. That's without eating, sleeping, or doing anything else. In reality, though, there are a few more positions than that. If a person wanted to see all possible orientations of the Rubik's Cube in that amount of time, he or she would need to enlist 43 billion of his or her closest friends—about six and a half times the current world population—to spend the next 33 years of their lives going through all the possibilities. We hope that gives you an idea of the 43,252,003,274,489,856,000 possible positions for the humble little cube.

Despite the vast number of positions the Rubik's Cube may begin at, any 3x3x3 cube can be solved in 27 or fewer turns. There are many different methods, but most people start by completing one face, then working either down or up the cube until all faces are complete. Some people compete at "speedcubing," or trying to solve a Rubik's Cube as quickly as possible. In May 2007 Thibaut Jacquinot set the world record by solving a cube in 9.86 seconds.

Build Your Own

Although making your own Rubik's Cube would be extremely difficult, you can certainly personalize one. Try replacing the solid-colored stickers on each side with a photograph. Cut each photograph to the size of the cube side, then cut it into nine "mini-cube"-sized squares. Glue them, in order, on the sides of the Rubik's Cube. When the cube is solved, it will display six complete pictures of your choosing, and when it's mixed up, no one but you will know what the images are.

Resources

If this 3-D logic puzzle is too easy for you, try a 4-D Rubik's Cube–type puzzle at the Magic Cube 4D Web site (www.superliminal.com/cube/cube.htm). Snoop around on the Internet—you may find a puzzle that features even more dimensions that are possible to represent only in a virtual environment.

SILLY PUTTY

History of Silly Putty

Created by accident, the material that would eventually be known as Silly Putty began its life as an invention without a use. In 1943 James Wright combined boric acid with silicone oil and produced a blob that bounced. General Electric, Wright's employer, sent samples of it to labs around the world, hoping that someone would find a practical application. No one did, but they all had a lot of fun playing with it. Several years passed before marketing whiz Peter Hodgson saw its potential as a toy. Although he was heavily in debt at the time, Hodgson bought a supply of the putty, and because Easter was coming up, he packaged it in plastic eggs. A year later an article in *The New Yorker* caused an explosion of interest in the toy and launched Hodgson to toy world stardom. Today Crayola owns the rights to Silly Putty.

According to the Web site Silly Putty University (www.sillyputty.com), more than 300 million eggs have been sold since the toy's 1950 launch. That's around 4,500 tons, enough to fill a Goodyear Blimp. (Imagine jumping onto that—it would be like bungee jumping without the cord!)

After it gained popularity as a toy, people did start to find practical uses for Silly Putty. It's been used to pick up dirt, lint, and pet hair and to stabilize wobbly furniture. But its crowning achievement took place in December 1965, when Apollo 8 astronauts Frank Borman, James Lovell, and William Anders used Silly Putty to secure tools while in space.

How Silly Putty Works

Viscosity is a measure of how well a liquid resists forces that cause it to flow. Water has low viscosity, while maple syrup has high viscosity. Over long periods, Silly Putty acts like a very viscous liquid, but applying a sudden force to it makes it act like an elastic solid similar to rubber. Place an upright blob of Silly Putty on a table and it will slowly flow into a puddle, but drop it onto the floor and it will bounce, and hit it with a hammer and it will shatter. Substances whose viscosity changes depending on the pressure that's applied are called "thixotropic." Other examples of thixotropic liquids are margarine (it doesn't flow until it's spread it with a knife) and latex paint (it flows under the pressure of a brush, but it doesn't run after the brush is removed).

For generations, kids knew that there were three things to do with Silly Putty: make a ball and bounce it, break it apart or mold it into different shapes, and use it to copy images from the Sunday comics. Alas, today we're down to two out of the three. Newspapers have changed the way they print the paper; they use different inks in much smaller quantities. Although these changes are great for saving money and natural resources, they make it difficult for the Silly Putty to pick up the images from the comics. But give it a try anyway. If you can copy Spider-Man, tug on his body and distort his head. Then roll him inside a ball to make him disappear.

In addition to the other odd features that Silly Putty exhibits, it also acts strangely when forced through a pipe or tube. Most fluids maintain their volume as they emerge from a pipe. But Silly Putty grows in volume. According to Jearl Walker's *The Flying Circus of Physics*, the molecules are stretched as they are forced through a pipe. When relieved of the pressure, they contract and swell up.

Inside Silly Putty

It's not too interesting inside Silly Putty—until you get down to the molecular level. Then we find that Silly Putty is made of 65 percent dimethyl siloxane (hydroxy-terminated polymers with boric acid); 17 percent silica, quartz crystalline; 9 percent thixotrol ST; 4 percent polydimethylsiloxane; 1 percent decamethyl cyclopentasiloxane; 1 percent glycerine; and 1 percent titanium dioxide. Nothing to it really; everything you would expect to find in a thixotropic like Silly Putty.

Make Your Own

1. Go to the grocery store and buy a pint of your favorite ice cream, some Elmer's white glue, and borax.
2. Eat the ice cream. Now you not only feel good, but you also have an empty container in which to hold your Silly Putty.
3. Pour ½ cup water into the ice cream container and add ½ cup glue. Mix thoroughly.
4. In a separate container (maybe you should have purchased two pints of ice cream!) mix ⅓ cup water with ½ teaspoon borax.
5. When you have mixed up each solution thoroughly, add the borax/water mixture to the diluted glue.
6. Stick your hands in the gooey mess and squeeze away to mix it thoroughly. Knead the mixture for several minutes.

Give the concoction a try: drop it, try to copy the comics with it, and tear it apart. Add more borax to make a better bouncing ball. The molecules in Elmer's glue are long polymers (polyvinyl acetate). Borax links the long molecules into a solid that flows.

SLINKY

History of the Slinky

Naval engineer Richard James invented the Slinky in 1943, and it has become one of the most popular toys of all times. Over 300 million have been sold. Richard's wife, Betty, came up with the name by thumbing through a dictionary looking for a word that sounded like the motion of the toy. The original price for a Slinky was $1. Now, more than 50 years later, the price is only a few dollars more, which means that when you adjust for inflation, a Slinky costs less today than it did when it was introduced to the world. The success of the Slinky spawned the Slinky Dog, the plastic Slinky, and other toys.

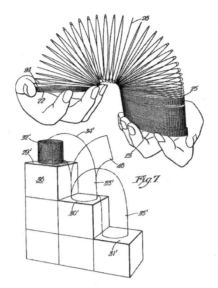

Patent no. 2,415,012

How Slinkys Work

Slinky consists of an 80-foot-long helical coil of metal or plastic. It does not demonstrate perpetual motion; it requires the potential energy that you provide by lifting it up the stairs to keep going. Push it off the top step and it converts its energy of position (potential energy) into energy of motion (kinetic energy)—gravity accelerates and pulls the Slinky down. (Slinky's motion also demonstrates Hooke's law, which states that the force required to stretch a spring is proportional to the length of its extension.) But Slinky doesn't stop after descending one step; it keeps going. Forward momentum (mass times velocity) carries the new top of the spring (the former bottom) over the edge of the step so it can fall down to the next lower level (and become the newest bottom).

Inside the Slinky

Inside the Slinky is air. A Slinky is all about the outside.

Science Experiments

Slinkys are used to demonstrate the two types of seismic wave in earthquakes. The P-wave (the P stands for pressure) is a compression wave; it is demonstrated by pushing and pulling a Slinky along its axis, from one end to the other. S-waves (the S stands for shear) are generated by moving one end of the Slinky from side to side, applying sideways shear to the coil. The two waves travel at different speeds in the earth and in the Slinky, which makes the demonstration even more appropriate.

If you twirl a Slinky around you while holding each end in a different hand, you can demonstrate centrifugal acceleration: as you spin faster the coil bulges farther away from you.

Build Your Own

Make your own futuristic music machine out of a Slinky. Poke a hole in the bottom of a Quaker Oats box or other cardboard, plastic, or metal cylinder. Insert one end of a Slinky into the hole. Use a wooden stick or metal bar to strike or rub the coil while listening with one ear near the opening of the cylinder. You'll hear some very cool sounds. To conserve these precious resources, cut a Slinky in half to make two music machines. Or you can put both ends in cylinders and play music in stereo.

Stilts and Jumping Stilts

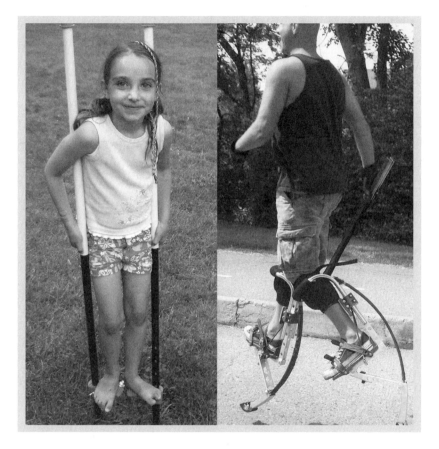

History of Stilts and Jumping Stilts

Stilts were developed for practical reasons and later adapted for fun. Shepherds used them to gain better views of their flocks and predators. Some people used them to cross streams or wetlands. Many cultures have developed and used stilts for a variety of reasons. Today, drywall installers and orchard crop harvesters use stilts. Sylvain Dornon was a pioneer in stilt walking; he walked on stilts from Paris to Moscow in 1891.

Jump now to the 21st century, when German inventor Alexander Bock invented a "device for facilitating the locomotion of a person" (U.S. patent number 6,719,671). The jumping stilt was born. There are several competing brands and squabbles over trademarks in this high-bounding business.

How Stilts and Jumping Stilts Work

The challenge with stilts is to keep your balance. In order to stay balanced you must keep your center of gravity above your point of contact with the ground. If you have more than one point of contact, your center of gravity must be above the area surrounded by your points of contact. Normally when you're standing you have many points of contact on the bottom of each foot. You can shift your weight from one foot to the other and from your heel to the ball of your foot or your toes. Shifting your weight allows you to stay balanced, usually without even thinking about it. With stilts you have only two points of contact— you can't shift your weight from back to front, and you can shift it only slightly from side to side. So when you're on stilts, you need a new method of balancing. You can achieve this by using a walking stick to give you a third point of balance, but most people think of this as cheating. The other way is to keep moving your feet. By doing this you are always falling over, but before you fall too far, you move a foot and catch your fall. This starts you falling in a new direction, so you have to keep moving to stay up.

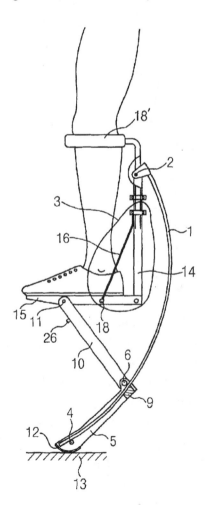

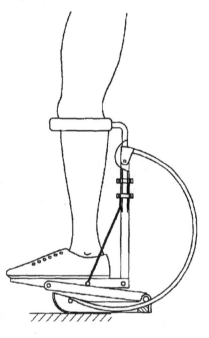

Patent no. 6,719,671

Powerisers, a brand of jumping stilts, work the same way as stilts when it comes to staying balanced. You still have only two points of contact, and you can't shift your weight forward and backward. The main difference between stilts and Powerisers is that Powerisers have springs on the back of them. These springs compress when you step on them, then release the energy, propelling you up again. They act like pole-vaulting poles to store energy briefly before redirecting it. When you plant your foot and stand on it, your kinetic energy—the energy of your foot moving—is used to compress the fiberglass spring on the back of the Powerisers. This transforms the kinetic energy into potential energy (stored energy). When you start to step off that foot, the potential energy is converted back into kinetic energy, which propels you up. By storing then releasing kinetic energy, you can jump much higher. There are many examples of kinetic energy and potential energy changing back and forth. Rubber band planes, trampolines, battery-powered cars, throwing a ball in the air, and even your muscles are all examples of kinetic and potential energy conversions.

Inside Stilts and Jumping Stilts

When it comes to stilts, there's not much to take apart. Traditional stilts basically consist of posts with foot pegs attached to them. Fancier stilts have straps to tie the stilts to your legs so you don't have to hold them, but that's about all there is to them.

Poweriser jumping stilts have a few more parts to them, but all are visible without taking the devices apart. A ski-like binding holds the foot in place, and straps secure the stilt to your leg. It takes some time to learn the balance, but Woody picked it up quickly and now bounces around the streets of Boise on a pair.

Build Your Own

Building stilts can be as easy or as hard a task as you want. The easiest way is to get two 8-foot-long two-by-twos and about 16 inches of two-by-four lumber. Start by cutting the two-by-four in half at a 45-degree angle, forming two trapezoids—the foot braces, or standing posts. Each trapezoid is attached to a two-by-two, with both trapezoids placed at equal distance from the ground. They can be attached permanently (with glue and wood screws) or be made adjustable. For adjustments, drill two holes through each foot brace 4 inches apart. Then drill holes at 4-inch intervals in the lower section of the two-by-two poles, starting about 6 inches from the bottom. The foot braces can then be attached with bolts and wing nuts.

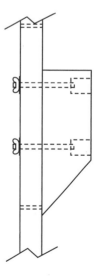

When you're ready to try them out, hold on to the poles, step onto the foot braces, and start walking. With every step you take, you'll have to pull up on a pole to raise the stilt off the ground. As you get better at this, you can raise the foot braces higher. If you get good and feel confident on the stilts, add some Velcro straps to the foot braces and poles to strap your legs in. Then you won't have to hold the stilts and your hands will be free for other things, such as juggling.

If you are skilled in metalworking, you might try making your own jumping stilts. Study a pair of Powerisers to understand the angles and how the springs are held in place. Double your medical coverage and give it a go.

STOMP ROCKET

History of the Stomp Rocket

The D&L Company made the first Stomp Rockets in the 1990s. It is based on a 1978 patent (patent number 4,076,006) by inventors Jeffery D. Breslow and Eugene Jaworski. This patent references a toy target air gun that was patented in 1912. The earlier patent envisions contests of skill by players who hit the top of a plunger that compresses a bulb, forcing a blast of air and a ball out the barrel toward the target.

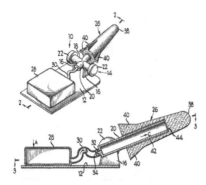
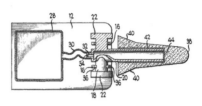

Patent no. 4,076,006

How Stomp Rockets Work

Although it's not really a rocket, it flies like one. A rocket, by definition, has an onboard engine and fuel, and it propels itself. This toy clearly does not meet that standard. However, in flight it looks like a rocket. Fins control its trajectory. Clip one off or add more fins to see the impact. The fins add drag to the back of the rocket, which keeps the rocket pointed forward. Putting fins on the front of the rocket will cause it to flip around in flight.

Fins and wings, which are often confused, have different functions. Wings are located near the center of a plane's fuselage and provide *lift* to keep the plane in the air. Fins are located at the rear of a rocket to provide the *drag* needed to keep the rocket moving in a relatively straight path. The addition of anything else that sticks out from the fuselage of the rocket body tends to add drag and slow the rocket.

Weight also matters. If the rocket is too light, it won't carry far. It leaves the launch pad at top speed, but quickly slows down and falls to the ground. If it's too heavy, it accelerates slowly and soon crashes.

The Stomp Rocket launcher chamber collapses under the compression of your foot and shoe. That forces air out of the rubber bladder into the plastic tube. The sudden blast of air hits the inside of the rocket and accelerates it skyward. The longer the rocket stays in contact with the launch tube, the longer it has to gain energy from the blast of air and the farther it will travel.

Inside the Stomp Rocket

There isn't much inside the launcher or the rocket. The rocket is made of foam and is hollow so it can fit onto the plastic launch tube. A clear plastic tube carries the blast of air that sends the rocket skyward. The tube has an unfortunate habit of collapsing, which decreases the air flow to the rocket. The bellows is made of tough rubber to withstand the heartiest of stomps. There are no valves; just a rubber bladder. After the user stomps on it and forces out most of the air, the bladder rebounds and draws air back through its one and only orifice.

Build Your Own

As much fun as these rocket sets are, you can make your own launcher and your own rockets that fly just as well. A simple launcher requires a two-liter plastic soda bottle, a bike inner tube, and 10-foot-long piece of ¾-inch PVC pipe. Use the inner tube to connect the bottle to the PVC. Tape one end of the inner tube to the bottle with masking tape; duct tape the other end to the PVC tube. You're ready to launch . . . once you've built your rocket.

More elaborate launchers include a protractor and a swivel launch tube mounted on a vertical pine board. The swivel allows rocket scientists to pick their launch angle, and the protractor allows them to measure and record information regarding which angles were most successful.

The coolest thing is that you can make your own rockets. Heavyweight paper is good, but discarded office paper works well, too. Roll a piece around a 10-inch length of ¾-inch PVC pipe, and loosen the paper so it slides on and off easily. Tape it together to make a cylinder. Fold in one end for a nose cone and make sure it's airtight. Add a paper clip or two to the nose.

At the other end of the rocket, add some fins. Business cards or index cards make great fin material; they are stiff and strong, but light. Most people prefer big, funky fins like those you'd see on a 1950s Cadillac, but they provide so much drag that they slow the rocket down. Fins need to be small, held securely to the rocket fuselage (the paper cylinder) with masking tape, and parallel to each other. If they are offset (parallel to each other but not to the fuselage), the rocket will spin in flight. The spin may provide stability to the rocket's flight, but the tradeoff is a bit of drag. Offset fins also slow down the rocket, much like a parachute on the back of a dragster slows down the racing vehicle.

The inexpensive materials required to make your own rocket allow you to try many different designs without spending more than the cost of a box of paper clips and a roll of masking tape.

Slide the rocket onto the launcher, then stomp in the center of the bottle to launch it. Use just one foot; if you use two at the same time, one is likely to land on the bottom of the bottle, irrevocably condemning it to the recycling bin. When operated properly, each bottle will yield about 50 launches.

Resources

For more Stomp Rocket possibilities, see Ed's books *Loco-Motion: Physics Models for the Classroom* (Zephyr Press, 2005), *Rocket-Powered Science* (Good Year Books, 2006), and *Inventing Toys* (Zephyr Press, 2002).

SUPER SOAKER

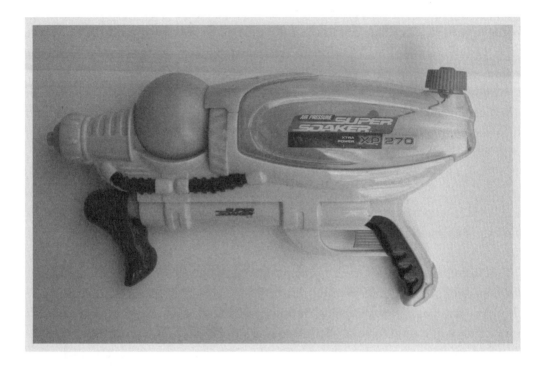

History of the Super Soaker

The Super Soaker, originally known as the Power Drencher, was invented by aerospace engineer Lonnie Johnson in 1988. It embodied a different concept for squirting water than that used in traditional water pistols. Rather than pumping individual squirts of water, it had you pump up pressure in a reservoir and use the stored air/water pressure to produce a far-reaching stream. Johnson got the Larami Company interested in his invention, and the sales success led to Hasbro buying Larami. Some 300 million Super Soakers have been sold, and 125 different models have been introduced to the market.

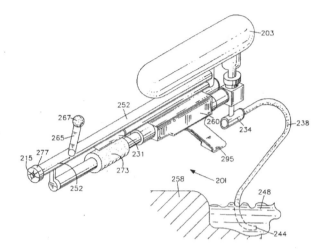

Patent no. 5,322,191

How Super Soakers Work

When you pull out the Super Soaker's piston a valve opens up, letting water flow from the fill tank into the pump. When the piston is pumped the valve closes and forces water past a second valve into the spherical (pressurized) reservoir. The more times you slide the pump, the higher the pressure in the reservoir and the farther the Super Soaker will shoot. Maximum pressure is about 35 PSI.

As the pressure builds in the reservoir, it gets harder to force more water in. At a particular pressure level a third valve opens, letting the water escape back into the fill tank. From there it vents into the air. This feature added additional complexity and more plumbing to the Super Soaker, but it prevents you from overpressurizing the reservoir.

Johnson later added a secondary reservoir to maintain pressure while the primary reservoir emptied. With this addition, the toy could shoot a stream of water 40 feet.

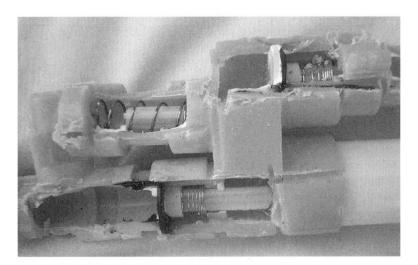

Inside the Super Soaker

To get inside the Super Soaker, start by making sure there is no pressure in the system. Pull the trigger and let air and water escape. Then remove the two rings that hold together the two halves of the gun. One is around the barrel; the other surrounds the piston/pump handle. Using a screwdriver, pry these forward, being careful not to spear your hand in the process. Once these are removed you can unscrew the several Phillips screws.

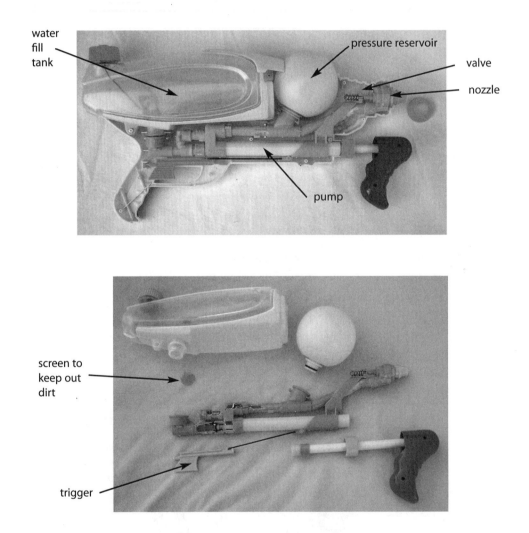

The trigger is linked to a valve at the end of the barrel. The ball is the reservoir. Follow the tubes to see how water gets into the ball and how it gets out again. To make it easier to see, add food coloring to the water and hold the mechanism up to the light to see. Remove the water reservoir and see that there is a metal screen in the pipe. This helps prevent debris from getting to the nozzle and clogging it.

Don't pressurize the gun when the plastic shell is removed. Pumping it up will cause the reservoir to fly off, potentially causing injury and certainly spraying water about. How did we figure out that out? Let's just say that experience leads to better judgment.

Build Your Own

You can build simple systems that mimic the Super Soaker. We used a ½-inch PVC pipe for the barrel, with an end cap into which we had drilled a small hole. For a trigger we used a ball valve. A PVC elbow joint connected the ½-inch pipe to a piece of ¾-inch pipe. We inserted a screw fitting so we could open the pipe to add water. Through the end cap on the ¾-inch pipe we drilled a hole large enough to insert a bike tire inflation stem, which we glued in place. Bike stores throw away used inner tubes all the time, and clerks there are usually happy to let you have discarded ones with still-functioning valve stems.

To operate our contraption, we unscrewed the pipe, filled it with water, and screwed it shut. Using a bike pump, we increased the pressure to 40 PSI. When the ball valve was opened, the pipe shot a stream of water more than 20 feet. This homemade version isn't as easy to operate or as safe as the Super Soaker, but it's fun to see how a simple system works.

VIDEO GAME LIGHT GUN

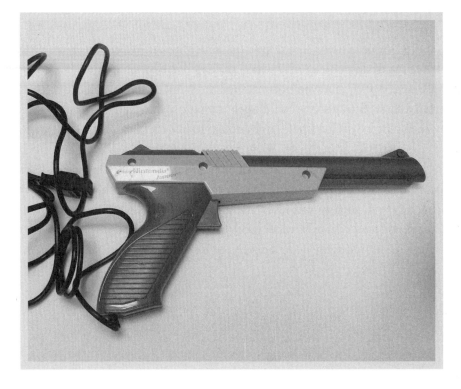

History of the Video Game Light Gun

When the NES game system was launched in 1985, it was bundled with Nintendo's version of the video game light gun, the NES Zapper, and a game called Duck Hunt. In this game, ducks would fly up from behind the brush, giving the player just a fraction of a second to sight in and shoot each one with the Zapper. The system was able to sense whether the Zapper's aim was true: a successful shot would dispatch the duck and increase the player's score, while a miss would cause the trusty onscreen hunting dog to snicker at his or her ineptitude.

About the same time, Sega released its Light Phaser as part of its Sega Master System. Both companies have created other versions of light guns to work with their newer game systems.

How Video Game Light Guns Work

The gun is plugged into the game controller and gets its power from the controller, but it turns out that it doesn't actually shoot anything. It doesn't send out a beam of light or sound, and it certainly doesn't launch birdshot at the ducks. But it does *look for* a flash of light.

With this model, when you pull the trigger you close the microswitch that sends a signal to the game controller. The trigger pulse causes the computer to blank the screen (for $\frac{1}{30}$ of a second), then highlight the target in white (for $\frac{1}{30}$ of a second). If you have an operating game system set up with a light gun, look to see the screen go blank when you pull the trigger. You might see this more easily if you turn off the room lights.

The optical sensor in the barrel uses a photodiode to detect light. Photodiodes pass electricity when exposed to light and block electricity when not lighted. When the photodiode in the barrel senses the change from a blank screen to a white target, the circuit board counts the shot a hit. If the gun is not aimed at a duck, it won't see the white target (after the screen blanks). You missed and the dog laughs.

Another approach to making a video game light gun is for the television screen to turn white after it blanks. Since the TV picture is created line by line, it takes time for the screen to fill with white. Even if this occurs in only $\frac{1}{30}$ of a second, in the world of electronics that is a lot of time. As the screen turns white, line by line, eventually the gun photodiode senses the white. It sends a signal to the game controller, which computes the screen position of the advancing white light. If that is also the position of a target, you hit it.

Inside the Video Game Light Gun

A half dozen or so Phillips screws hold the two halves of the light gun together. Remove them and lift off the left half of the plastic shell. Be careful to set the right half gently on a table—otherwise its parts will fall out.

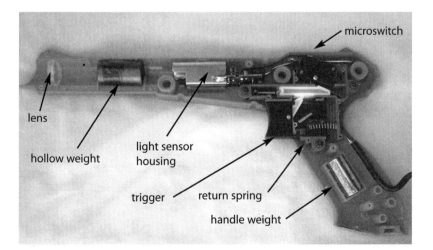

Near the end of the barrel you'll see a small, clear plastic lens that keeps dust out of the gun and focuses light on the optical sensor (a photodiode) behind it. The next object in the barrel is a hollow weight. It's hollow so that light can pass through. This weight and the shiny weight in the handle are there to give the gun the right heft, or feel. Without the weights, the gun would feel like an empty water pistol.

Behind the hollow weight is a shield that covers the optical sensor and small circuit board. Below this apparatus is the trigger assembly, which you can remove by taking out three screws. The trigger assembly of our model has three springs. The spring on the bottom pushes the trigger back to its normal position after it's been pulled. The one on top pulls the sliding lever back to its normal position. The third pulls the catch mechanism back.

When you pull the trigger, you slide the lever in the trigger assembly to the rear. That pushes up on the metal lever, which pushes to close the switch. The black rectangular device with the lever and switch is a microswitch.

Resources

Video game light guns turn up at thrift stores and garage sales frequently. Usually costing about a buck, they're a bargain to take apart.

VIEW-MASTER

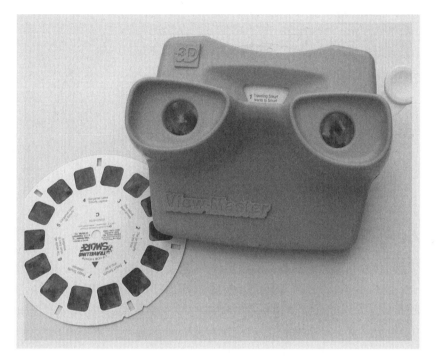

History of the View-Master

When organ maker and 3-D photographer William Gruber met Harold Graves, president of a photo company, in 1938, things were bound to click. Both were visiting the Oregon Caves National Monument, where Gruber was taking stereo (3-D) photos with two cameras mounted side-by-side on a stand. They discovered their mutual interest, and that led them to debut the View-Master in 1939. The new product was sold by Graves's picture postcard company, Sawyer's, Inc., of Portland, Oregon. They marketed the viewer as an improvement on the picture postcard: you got several images, and all were in stereo. Sales really took off when the U.S. military bought 100,000 viewers for training during World War II. After the war, children became the primary customer base.

The original View-Master was made of Bakelite, the first plastic. Later, thermoplastics (plastics that melt when heated) were substituted. The View-Master is still produced, although in 2000 production was moved from Portland to Mexico.

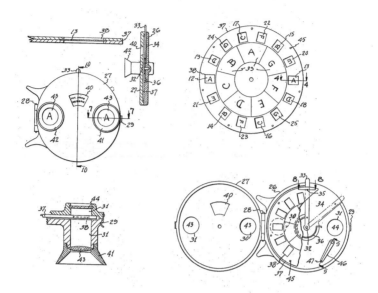

Patent no. 2,189,285

How View-Masters Work

In order to see a 3-D image, your two eyes have to each see slightly different things. To mimic the separation of your eyes, 3-D cameras are set a few inches apart. That way, each camera sees the objects to be photographed from slightly different angles, just as your eyes do.

When you look at a 3-D movie or picture, you use a viewer that allows your left eye to see only the left image and your right eye to see only the right image. For screenings of 3-D movies, viewers are given goggles that block their eyes from seeing the other eye's images. The traditional way to do this is via different-colored filters for each eye or different polarization of the two lenses. Some modern systems have goggles that open and close lenses for the right and left eyes in sync with the projection of left and right images on the screen.

With the View-Master, you peer through binocular lenses at stereo images mounted in cardboard disks. The lenses allow your left eye to see only the left-hand image, and your right eye to see only the right-hand image. The images are transparencies (slides) that require you to hold them up to light to see. Translucent material at the distant side of the View-Master lets in the required light.

Inside the View-Master

Four plastic rivets hold the front and back halves of the View-Master together. Pry a flat head screwdriver between them to pop it open. Hold the eyepiece up to your eyes and look at a finger; it will look fuzzy until you move it close to the viewer. Check out the apparent size of your finger; the lens magnifies images to about twice their actual size.

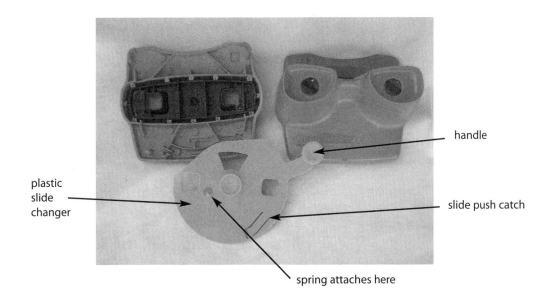

plastic
slide
changer

handle

slide push catch

spring attaches here

Nestled between the two halves of the viewer is the plastic slide changer. When you depress the handle, a latch molded into the slide changer catches in the photo disc and rotates it. When you release the handle, a spring returns the handle and the latch slides past the photo disc without engaging it.

Build Your Own

You can create your own 3-D photos by renting a 3-D camera from a photo store or by mounting two cameras on a platform or bar so that they are held in place with their lenses about one eye width apart. Aim each camera at the same object and take pictures with both cameras at the same time. To view them, block vision between your eyes with a piece of cardboard so that each eye sees only one image.

Vortex Tornado

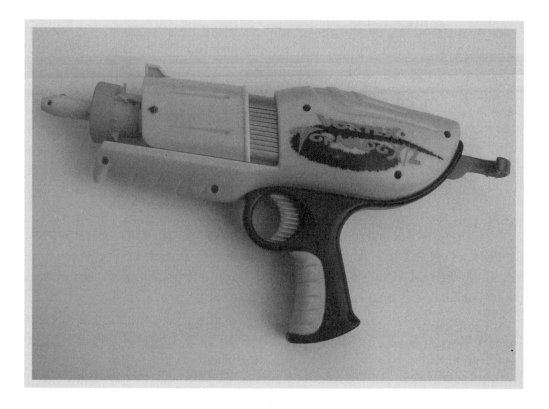

History of the Vortex Tornado

Chester F. Vanek and seven other inventors were awarded patent number 5,970,970 in 1999 for this "ring airfoil launcher." OddzOn Products, Inc., brought the toy to market. OddzOn got its start by inventing the Koosh Ball in 1986.

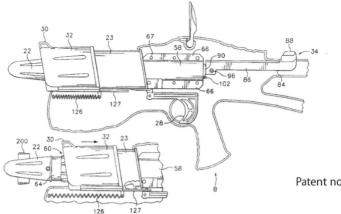

Patent no. 5,970,970

How Vortex Tornados Work

This spring-powered toy propels small rubber rings much farther than the advertised 60 feet. The user provides the energy for the spring by pulling back on the "T" handle at the rear. Pulling the trigger releases the slide, which moves forward and rotates under the force of the main spring and the guidance of a slot cut in the side of the barrel. The slide imparts both forward motion and spin to the rubber ring, which will fly clear into the neighbor's yard.

The ring flies in a straight line; it doesn't rise in the air like most thrown or shot projectiles do. The level flight, stabilized by the spin imparted by the launcher, helps the ring to travel far. In this way, the Vortex Tornado resembles the X-zylo (see page 168), another ring airfoil that flies level and far.

Inside the Vortex Tornado

Crank up the screwdriver—there are quite a few Phillips screws here. Remove them all, including the one found underneath the rubber ring on the barrel. Then you can lift off half of the case.

When you load a ring into the launcher you first pull back the cover and hold it in place by squeezing the trigger. Pull the cover rearward now to see how the trigger uses friction to hold it. Release the trigger and the cover flies forward, drawn by the spring underneath the cover.

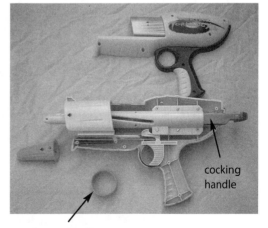

cocking handle

ring projectile

slide follows slot   spring that   peg that fits into slot
                        propels the ring

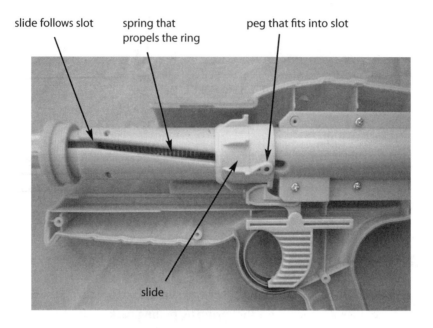

slide

To cock the launcher you pull back on the "T" handle at the rear of the launcher. This elongates the main spring, which is anchored at the front of the launcher and is attached to the slide. Notice that as you pull the handle back, the slide moves to the rear and rotates. Two pegs on the slide fit into curved slots in the firing mechanism. A hemispherical piece of plastic on the slide fits into the notch on the trigger to hold the slide in position. When you pull the trigger the slide escapes from it, moves forward rapidly, and rotates. The impulse from the slide spins the rubber rings. The rubber ring on the barrel absorbs the shock of the spring coming forward and slamming into the stopped position.

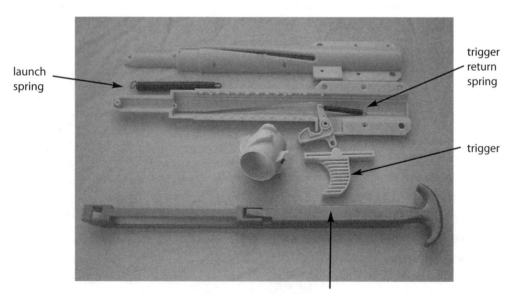

launch
spring

trigger
return
spring

trigger

handle for cocking spring

In order to go beyond this point in your disassembly, you're going to have to do some damage to the launcher. You can remove the small screws holding together the two halves of the interior mechanism. However, to separate the two halves you have to get the slide to move forward and off the barrel. The shiny pin in the slide that anchors the spring prevents this. Before you knock out the pin, pry the halves far enough apart to release the main spring from its anchor. That way, it won't "boing" you when you punch out the pin. Next, cut off the head of the pin, then use a hammer and nail to punch out what's left of it.

Once the device is completely disassembled, you can see that there's another spring that provides return force for the trigger. To reassemble your Vortex Tornado, you'll have to replace the pin and glue it into position.

Build Your Own

It'd be tough to build your own version of the Vortex Tornado launcher, but the X-zylo launcher (see page 170) is similar and easier to recreate. You can also mimic the flight characteristics of the rubber rings by folding paper into a cylinder. Even better is cutting the top and bottom off a disposable one-liter plastic water bottle, leaving a cylinder slightly longer than its diameter. Glue a discarded electrical cord (from an old appliance) to the leading edge of the cylinder to form the lip. Hold the ring like a football (with lip forward) and give it spin as you fling it.

Resources

Ed's *Fantastic Flying Fun with Science* (McGraw-Hill, 2000) shows in detail how to make the paper and plastic flying cylinders.

WATER PISTOL

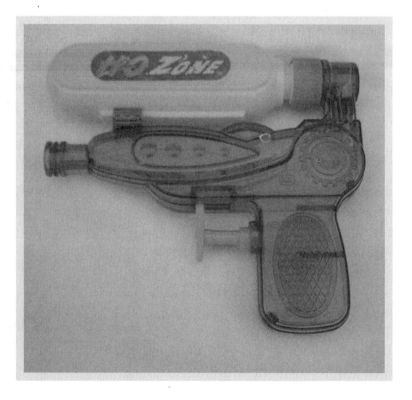

History of the Water Pistol

The earliest U.S. patent we found for a water pistol was issued in 1896. John Wolff invented a gun that uses a bulb to store and squirt water. According to his design, by placing the barrel in water and pressing the trigger once or twice, water is sucked into the bulb. Pressing the trigger again sends a squirt of water out the barrel. Since that patent was issued, several hundred types of water pistols and guns have been invented.

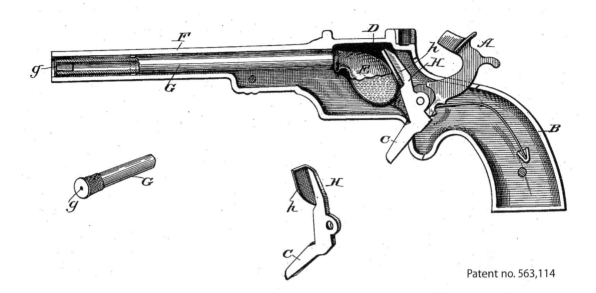

Patent no. 563,114

How Water Pistols Work

A water pistol is a pump. It draws water from a reservoir and forces it out through a narrow opening. You operate the pump by pulling on the trigger. To keep water from sliding back into the reservoir between each trigger pull, the gun has check valves that open and close with every stroke. These are simple devices—usually a ball or another piece of rubber or plastic that blocks water flow in one direction.

It takes a few pulls of the trigger to prime the pump: you must force air through the pump to draw water up through the bottom tube before you begin to pump water. Once water is in the pump, each trigger pull forces water past the top valve and out the tube. Since the nozzle opening is small, the water comes out in a stream.

Relaxing the trigger allows the spring to push the trigger back. Reduced pressure lets the top valve shut and the bottom valve open, admitting water. On the next squeeze of the trigger, the increased pressure closes the bottom valve so water can't escape back into the body of the gun, and opens the top valve so water can move toward the nozzle.

Inside the Water Pistol

The plastic body is a reservoir for water. To get inside it, we used a hacksaw and a rotary cutting tool. The interesting parts of this toy are the pump and the nozzle. The opening in the nozzle is tiny, even compared to the size of the tube that feeds it.

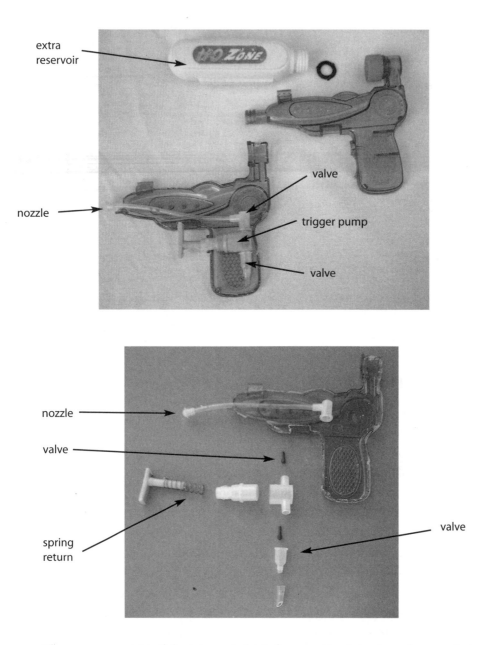

extra
reservoir

valve

nozzle

trigger pump

valve

nozzle

valve

spring
return

valve

The pump consists of the trigger (which forces water into the exhaust tube), valves, and connecting tubes. Water is sucked into the pump through the tube at the bottom. There is a spring inside the pump to return the trigger after it's squeezed.

If you're really in the mood to take stuff apart, check out the pump that dispenses hand cream and similar products. To get it open you'll need a strong knife or other cutting tool (and maybe a box of adhesive bandages). It's worth the effort to see the engineering of these small pumps.

Build Your Own

This doesn't look much like a water pistol, but it does pump water and squirt it a few inches. Purchase a baster—one of those kitchen gadgets used only at Thanksgiving to baste the bird. (Don't use one you want to keep, as you're going to destroy it.)

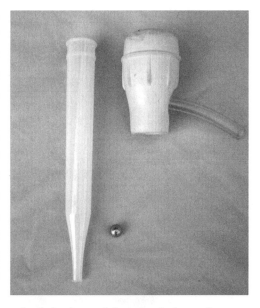

Take the baster to a hardware store and buy a ball bearing that's just small enough to fit inside it. Also get a few inches of plastic tubing that has an inside diameter of ¼ inch or less. Remove the bulb of the baster, use an awl to punch a hole in the bulb, and jam one end of the tube into it. Place the ball bearing inside the baster and replace the bulb, and voilà—you've made a pump.

Put the suction side of the pump (the end of the baster) into a pitcher of water. Squeeze the bulb, cover the end of the plastic tubing with a finger, and release the bulb. As the bulb regains its shape it will suck water up from the pitcher. The ball bearing will rise to let the water in, but will fall down into the narrow part of the baster to block water from leaving.

Remove your finger each time you squeeze the bulb so that air can vent through the plastic tubing. Cover the tube as you release the bulb. With each squeeze you'll force air out and with each relaxation you'll draw water in. Water will rise until it enters the bulb. On the next squeeze water will shoot out the tube.

OK, you won't be able sneak up on someone and give him or her an unsuspected squirt, but it's a neat pump nonetheless.

WATER ROCKET

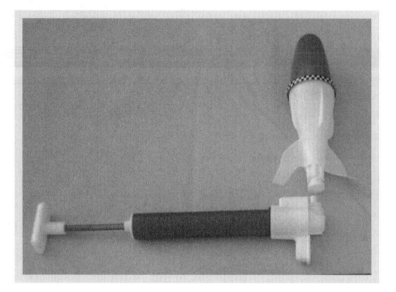

History of the Water Rocket

The idea of using plastic soda bottles as water rockets took off with the publication of an article in *Mother Earth News* in 1983. As often happens in invention, the article's authors were working on a different problem and their curiosity got the better of them. A few experiments later, and the results would inspire science fair projects and Science Olympiad competitions for years to come.

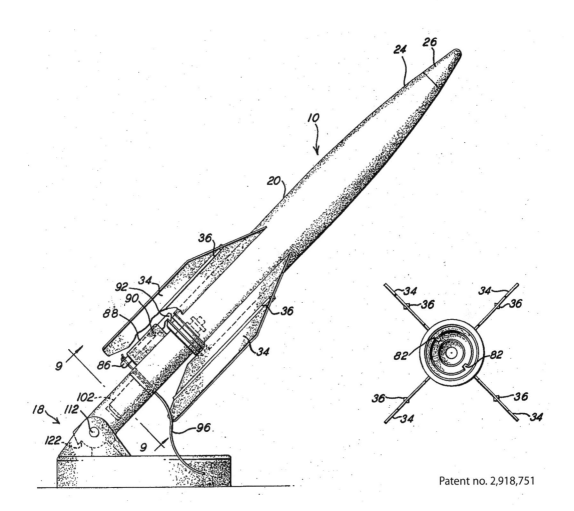

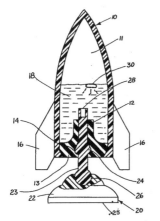

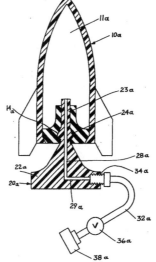

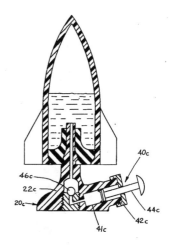

Patent no. 2,918,751

Patent no. 3,046,694

How Water Rockets Work

By pumping air into the rocket, your muscular energy is transformed into pressure, which upon release converts to kinetic energy (energy of motion). The pump sucks in air through the small hole located at its back (the same hole is used to lubricate the pump). The air you push forward in the pump goes into the rocket by pushing the steel ball up inside the valve. At some level of pressure, you won't be able to force more air into the rocket. Pull back on the release slide and the rocket launches.

On launching, the pressurized air and water inside forces water out the opening. The downward force of the water pushes the rocket upward.

So what role does the water play? Try firing the rocket without water. It will barely lift off. If you completely fill the rocket with water and leave no room for air, it will give a similar result. There is clearly an optimum level of water.

Air pushes the water out of the rocket, but the water (which has a density, or mass per volume, that's a thousand times greater than that of air) provides the change in momentum (mass times velocity) that propels the rocket. Air is compressible, so it is able to store the energy to fly the rocket.

Inside the Water Rocket

The rocket itself is a hollow plastic vessel. The other part, the pump, is neater. To get inside, we cut through the pump housing at each end. The handle fell out; it's connected to a piston and plunger that moves air. In the base that holds the piston is a hole to lubricate the pump.

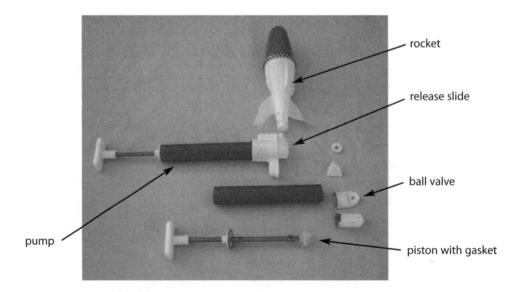

rocket

release slide

ball valve

pump

piston with gasket

The catch that holds the rocket to the launcher slides off the tube when you cut the other end off. In operation, it presses the base of the rocket against a rubber gasket, preventing air and water from escaping. Once the rocket is pressurized, you slide the catch along the handle to release the rocket.

What's not obvious is what keeps the air in the rocket. As you stroke the pump, you force air into the rocket. Why doesn't it come back out while you're pulling the handle back?

Look down the plastic tube that fits into the rocket. Inside you see a tiny steel ball. But there's more to the valve than just the ball. Cut on the seam just below the washer and get ready to catch the ball.

Look inside the chamber where the ball sits. When the ball is forced down by pressure in the rocket, it fits snugly against the housing to prevent pressure from escaping. But when high pressure in the pump forces it up, air can move from the pump, past the ball, and into the rocket because the chamber has indentations molded into its sides.

Build Your Own

You can purchase components from science catalogs to build your own version of this device, or you can build one from materials you purchase at a hardware store. A simple system requires a two-liter soda bottle, a bike pump, an inflating needle, and a one-hole rubber stopper. Glue the needle into the stopper, but be careful not to glue shut the air port. Fill the bottle about one-quarter full of water and insert the stopper. Rest the bottle

against an inclined board or brick, oriented so that the stopper end is down. Pump. As pressure increases inside the bottle, eventually it will force the stopper out and the bottle will fly upward.

The obvious problems with this design are that you can't control when the rocket will launch and you can't control how much pressure you pump into it before it does. It doesn't take much pressure to force the stopper out, so the rocket won't fly very high. Even so, don't try this indoors.

To control the time of launch—and pump up the pressure—you need to hold the neck of the bottle in place until you're ready to release it. One way to do this is to use two metal bars that clamp onto the neck just below the plastic ring of the two-liter bottle. See Ed's book *Fantastic Flying Fun with Science* for complete plans. Less expensive and easier is to order a launcher from a science catalog. See Resources, below.

Science Experiments

Water rockets offer the chance to perform dozens of science experiments, provided you can measure the pressure (use a pump with a built-in gauge) and launch height or time aloft. Determine the optimal level of water and air for maximum height. Evaluate the impact of different fin designs. Add weight to the nose and compare height of launch to the rocket's weight.

Resources

Plastic water rockets are sold at toy stores. You can purchase more durable and controllable launchers from the Science Source (www.thesciencesource.com; 800-299-5469), Edmund Scientifics (www.scientificsonline.com; 800-728-6999), or Nasco (www.enasco.com; 800-558-9595).

Full plans for a water rocket launcher are in Ed's book *Fantastic Flying Fun with Science* (McGraw-Hill, 2000).

WIFFLE BALL

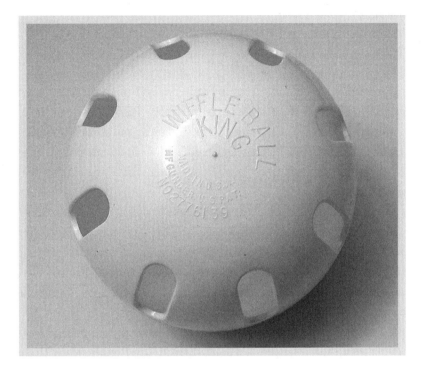

History of the Wiffle Ball

David N. Mullany invented the Wiffle ball in 1953 to help his son throw curves without wearing out his arm. He got some plastic balls of about the right size from a local cosmetics factory. He cut them in half, cut slots in one of the halves, then taped the two halves back together. He thought that by cutting holes in one section, the weight would be asymmetric and the ball would curve. He discovered that the weight of the object had little to do with its tendency to curve, but that properly spaced slots changed the aerodynamic forces on the ball. Father and son experimented to find the optimum number and size of slots and agreed on eight oblong slots. With the ball perfected, Mullany borrowed money to launch a family business that continues today.

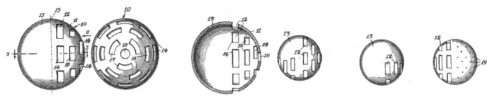

Patent no. 2,776,139

How Wiffle Balls Work

A baseball curves because the spin put on it creates an asymmetry. Air flows over one side of a spinning ball differently than it does over the other side. One side of the ball has forward spin and encounters a "headwind" from the linear motion of the ball as it heads toward the plate. The other side experiences the same linear motion as the ball heads toward the plate, but its spin is going in the same direction as the air it encounters, giving it a "tailwind."

Everyone can agree on that much of the explanation. But now you have to state your allegiance to either mathematician Daniel Bernoulli or to the rebel alliance. Bernoulli followers shout "Magnus effect!": the side of the ball with greater velocity (the sum of spin and linear motion) experiences lower pressure and the ball moves in that direction.

But the anti-Bernoulli forces claim that Bernoulli's principle (an increase in fluid speed accompanies a decrease in pressure), postulated for closed pipes, doesn't apply. Clearly there are no closed pipes at play here. The curve is caused instead by the Coanda effect: a moving fluid stays attached to a solid surface. The fluid bends with the contours of the surface. Thus, an airplane wing gives lift because air follows the curve of the wing downward. In doing so, the air imparts an equal and opposite upward force on the wing. Lift! No need to have streamlines meeting and molecules catching up to their buddies, as Bernoulli would have us believe.

In the case of a curve ball, the forward spinning side of the ball encounters the headwind and loses energy. Air flow separates from the ball sooner than it does on the backward-spinning side. The ball displaces air toward the slower (forward-spinning) side, so the air displaces the ball toward the backward spinning side. Strike one!

Anyway, you can throw a curve with a slotless ball. (Do it well and collect a few million dollars every year.) Throwing a curve ball using a very light ball is more difficult. The Wiffle ball is much lighter than a baseball, and it doesn't have enough angular momentum to keep spinning without help from an additional force. The slots provide the additional force of increased drag. As the ball is pitched, air moves into and out of the slots. This action slows that side of the ball and causes the ball to curve toward that side.

The light weight of the ball and its high drag cause it to slow quickly in flight. A hit ball doesn't travel far. The other key feature of the Wiffle ball—that it usually doesn't damage what it hits—is also attributed to its light weight, as well as to its somewhat flexible plastic construction.

Students at Harvey Mudd College have conducted wind tunnel experiments with Wiffle balls and have measured the forces of drag and lift. But more work on this important topic is needed. Perhaps NASA should jump in.

Science Experiments

To see the Coanda effect, tape a ping pong ball to the end of a foot of string. Insert the ball into a stream of water flowing from a faucet. Lightly pull on the string to see that the water is pulling on the ball by bending around it instead of falling straight down. This pulling force is the Coanda effect.

It would make a cool experiment to measure the effect of the slots on a purchased Wiffle ball. Cover some of the slots with duct tape, throw the ball at a target, and measure how far it curves to the right or left. Compare that to the curve distance achieved with all or none of the slots covered. To throw consistently, you'd need a pitching machine; a possible alternative would be to roll the ball down a ramp and off a tall building. That would make the curve distance easy to determine—measure from where the ball lands to the point directly under the ramp. If you try this, wear a hard hat and let us know the results.

Resources

Confused by all our anti-Bernoulli rantings? Check out *Understanding Flight* by David F. Anderson and Scott Eberhardt (McGraw-Hill Professional, 2001).

WINDUP TOY

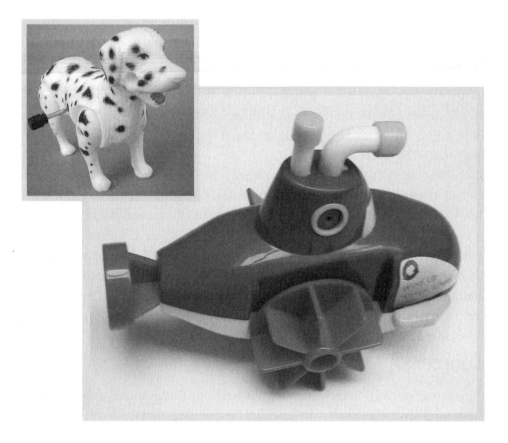

History of the Windup Toy

Spring-operated toys were popular throughout the 19th century. A toy velocipede was invented in 1877 by J. E. Conklin (patent number 191,278). The toy locomotive on the next page was patented in 1882 (patent number 267,939) as an improvement to spring-driven toys.

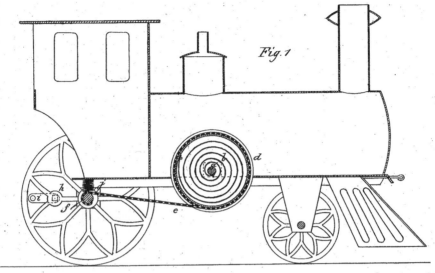

Fig. 1

Patent no. 267,939

How Windup Toys Work

When you wind up a toy you store energy in its spring. When the toy unwinds, the spring drives a set of gears.

Springs and other elastic materials obey Hooke's law. The law states that the amount of deformation in an elastic material is directly related to the strain or the force imposed. The farther you want to stretch a rubber band or spring, the more force you have to exert. The farther you stretch it, the greater the force available when you release it.

In the dog pictured on the previous page, the gears connect to two cams, each of which moves a flat plastic piece that conveys the motion to the legs. In the submarine, you wind up the internal spring by twisting the side-wheel paddles.

Inside the Windup Toy

Our windup dog is held together by three screws. By removing them and prying open the case, we were able to get to the inner devices.

This dalmatian has some neat parts. One windup motor operates all four legs and the tail. The spring and gears are encased in plastic, but the outer end of the spring is visible sticking out of the case. On each side of the drive axle is a cam. As the cam rotates it pushes a flat plastic piece, back and forth, up and down. Each leg connects to one of the two

Large gears drive small gears to speed up the rotation

flat plastic pieces and receives its motion from it. The leg movement on one side also pushes the tail up and down.

Without opening up the plastic case we can see a gear on the shaft that holds the spring. As we turn the windup knob, the spring winds, but the gear doesn't move.

The dog is kaput when we open the plastic case. Rover will rove no more. Prying open the plastic case, we can see how the gears disengage when the toy is wound. A ratchet on the spring shaft engages the first gear only when the spring is unwinding. When you wind the spring, the ratchet's teeth slip by the gears, making the clicking sound you hear while winding.

windup knob

spring

The spring is a piece of flat metal coiled up inside the plastic case. The windup mechanism connects to the inner end of the spring. The outer end is held in place by the case. The gears are arranged to speed up the rotation so that the final gear spins faster than the spring turns its axle.

The legs are attached like a first-class lever, in which the pivot point is found between the load and the effort. In this case, the pivot point is close to the effort so the lever magnifies the motion. That is, the flat plastic part moves a short distance to move the feet a much greater distance.

The dissection of the windup submarine requires only the removal of a few screws. The motor consists of gears and a spring.

Build Your Own

Rubber bands are much easier to use than springs to store energy to propel vehicles or other toys. We make rubber band launchers for paper airplanes by adding a reinforced hook to the bottom of a plane and making a launcher by looping a rubber band around the end of a dowel.

Rubber band boats are easy to make. Use a paper milk or juice carton for the hull. Make a paddle by cutting two rectangles out of extra pieces of the carton and cutting a slot in the middle of each. Slide them together at right angles and glue. For a stern-wheeler, extend

two dowels behind the boat and support the paddle with a rubber band looped around the ends of the dowel. For a side-wheeler, make two paddles and glue them onto the ends of a dowel that passes through holes in the sides of the carton. Wrap a rubber band around this dowel and tie it onto the bow or stern of the boat.

Both of these models could be made using springs. Check out the supply of springs at one of those hardware stores that have everything.

X-ZYLO

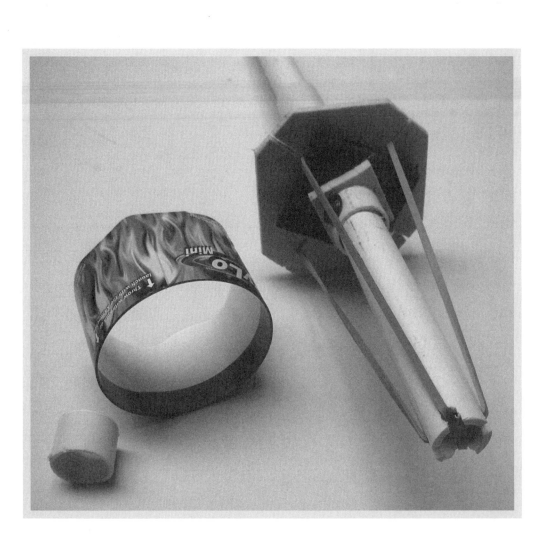

History of the X-zylo

Mark Forti, a college student at Baylor University, was flinging paper models across his dorm room when he came up with the design for this flying cylinder of paper. He formed a company—William Mark Corporation—with his father (William) and turned an idea into a booming business. Other companies make similar products.

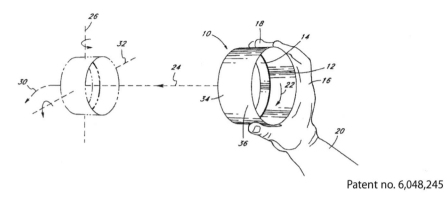

Patent no. 6,048,245

How X-zylos Work

We're not sure how the X-zylo flies so far—a casual launch will send it 65 to 100 feet—and neither, it seems, is anyone else.

Spin matters. It provides stability in flight. It also seems to generate lift. But since the X-zylo will fly even if you don't give it any spin (although it won't fly well) the cylinder itself must generate some lift. Air flying past the lip is partly responsible, but so, too, we surmise, is air flowing through the cylinder. We need an NSF or NASA grant to study this!

Inside the X-zylo

Put away the knife and screwdriver—you need only your eyeballs to peer inside an X-zylo. The toy is an open cylinder. One end is heavier than the other; this is the leading edge. On some models, the trailing edge often is cut into a wavy pattern. Most are made of plastic or plastic-covered paper.

Build Your Own

We've made similar toys out of paper, aluminum soda cans, and plastic water bottles.

Paper first. Use discarded office paper or heavier paper. Working from the side (lengthwise), fold the bottom third upward and crease it well. It's important to be neat and to form good creases. We use the always handy Swiss Army knife to crease each fold.

Now take the bottom edge and fold it so that it lies on top of what was formerly the edge of the paper. Repeat this step so that you make several folds, all ending about a third of the way up the paper. When you can't fold it any more, crease it well.

With a hand at each end of the paper, roll the ends into a cylinder. Interlock the folds from opposite ends together and tape the cylinder with masking tape. You can try giving it a fling now, but to fly well it will need some large paper clips inserted along the leading edge.

To fling your toy, hold it like a football between the thumb and forefinger of your throwing hand, with the heavier edge facing forward. As you bring your arm forward, roll the toy off your fingers to give it spin. Fetch it, adjust your throw or add more paper clips, and try again.

For a plastic X-zylo-type toy that's almost as good as the ones you buy in sporting goods stores, cut the top and bottom off a disposable one-liter plastic water bottle. The length of the remaining cylinder should be just a bit longer than its diameter, although this isn't critical. Add some weight (paper clips, or wire from a discarded appliance) to the leading edge; glue it and tape over it to hold it in place. And prepare to be amazed. Ready? Fling it!

But there's more! As cool as X-zylos are, you can make them even cooler by creating a launcher (or you can purchase one from the William Mark Corporation). One of the cool features of the launcher is that you can easily change how much spin you give the X-zylo to see the effects of spin on flight. As you pull the X-zylo away from the end of the launcher, you stretch rubber bands. You can twist the rubber bands to give spin in either direction, or you can not twist them and give no spin.

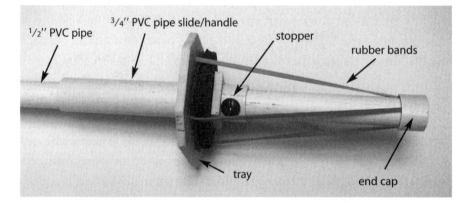

We made a launcher out of PVC pipe. A 30-inch-long piece of ½-inch PVC forms the launch guide. Three pieces of ¾-inch PVC form the handle, the stopper, and the end cap that holds rubber bands in place. We cut a 3½-inch square out of a ¼-inch PVC plastic sheet. We cut off the corners to make it an eight-sided piece, and we cut slots in each of the new edges to accommodate rubber bands. We bolted another piece of ¾-inch PVC to stop the forward motion of the handle and the X-zylo.

Next we used four 7-inch rubber bands for propulsion and held them in place by cutting slots in the end of the launch guide. A piece of ¾-inch PVC was forced over the end (and rubber bands) to keep them from flying off.

An X-zylo-type toy made from a piece of standard paper just fits onto the launcher. So do the mini X-zylos that the William Mark Corporation sells.

Resources

Check out the William Mark Corporation Web site (www.flyingproducts.com).

Yo-Yo

History of the Yo-Yo

There is quite a bit of mystery surrounding the origin and use of the yo-yo. There is some thought that the ancient Chinese had a toy that somewhat resembled a yo-yo as early as 3,000 years ago. Apparently, some yo-yos from 2,500 years ago have been recovered in Greece. Fast forwarding to 500 years ago, Philippine hunters began using a device that had wooden discs connected to a string. It is claimed they would throw this at prey and either hit the prey or entangle it in the string.

More certain is the yo-yo's recent history. In 1866 a patent was issued to two Americans for a device they called whirligig. Similar designs had been popular in Europe before that. But the toy didn't catch on in the United States until an immigrant busboy named Pedro Flores began demonstrating the toy in the 1920s. He created a market for the toy and made a few to sell to people. Business got stronger and he started a factory to pro-

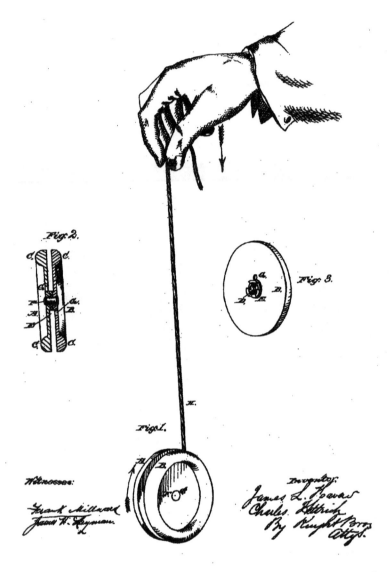

Patent no. 59,745

duce them. Donald Duncan saw the toy's potential, bought Flores's factory and other assets in 1929, and launched a toy revolution with the Duncan Yo-Yo.

Not only was Duncan a toy visionary, but he developed new ways to market them as well. One deal required kids entering a yo-yo contest to sell newspaper subscriptions; in return, the newspaper provided Duncan with free advertising. He sent yo-yo demonstrators onto school grounds to show kids how cool the toy was. And he was one of the first to use the then-new medium of television advertising to promote a product.

The Duncan Yo-Yo has been inducted into the National Toy Hall of Fame at the Strong National Museum of Play.

"*The parochial grade school I attended in the 1960s banned the yo-yo. To devote time to learning yo-yo tricks was to acquire a slightly wayward skill. A yo-yo marked out its owner as an idler who'd been lured astray. The Duncan slogan, a mantra almost, may have sounded like a rival liturgy, too: 'If it isn't a Duncan, it isn't a yo-yo. If it isn't a Duncan, it isn't a yo-yo.' Playing with the yo-yo took on the flavor of an outlawed pleasure. During the baby boom, playgrounds were parking lots, and during recess in these years the crowded asphalt, like kids' play itself, became mostly kids' territory. But still, the authorities could invade and confiscate a yo-yo if they discovered one. Once seized, the toy would be whirled and flung onto the school's flat roof, ritually sacrificed. (In memory, this smooth move has merged with martial-arts movie choreography; it's just not possible now to retrieve the moving image of the swirling black-and-white habit of the Dominican nun without also thinking of the kung-fu robes of the Shaolin priest.) We imagined that a shining surface, festooned with gleaming transparent Imperials, waited above. The roof would be ankle deep in multicolored Mardi Gras, sleek black Tournaments, and wooden Satellites that whistled. It was a field of dreams. We schemed at length and in vain for a path to the unreachable.*"

—Scott Eberle, Vice President for Interpretation, Strong National Museum of Play, National Toy Hall of Fame

How Yo-Yos Work

Throwing a yo-yo gives it kinetic energy of linear motion and rotational motion, as well as potential energy if the string is held from above. The fast-spinning discs have gyroscopic stability, so the yo-yo doesn't turn around the vertical string.

The yo-yo that paid for Pedro Flores's early retirement was fundamentally different from earlier designs. On earlier models, the string was tied securely to the axle or shaft. When the string had played out, the two spinning discs started to climb back up the string, converting some of the kinetic energy back into potential energy. Without a flip of the wrist, the yo-yo wouldn't make it all the way back into the hand, as friction converted some of the kinetic energy into heat.

But things got much more complicated with Flores's model. He tied the string onto the axle with a loop, allowing the axle to rotate inside the loop. The yo-yo could "sleep," or spin around without climbing back up the string. A sharp snap of the wrist would add just enough additional friction for the string to bind on the discs and climb up the string.

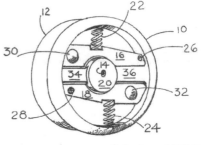

A more recent innovation has been to add ball bearings to reduce the friction. The balls are held in place between two circular steel tracks or races: one on the inside and one on the outside. The string is tied to the outer race instead of to the axle. When the yo-yo is sleeping, the outer race spins and the inner race, which is connected

Patent no. 4,332,102

to the axle, does not. A sideways jerk causes enough friction for the spinning balls to transfer energy to the inner race and get the yo-yo to rise.

Innovators never stop, and the next design enhancement would involve a centrifugal clutch. When the yo-yo spins quickly at the bottom of its throw, the centrifugal clutch disengages the discs from the axle. The discs can spin without the axle spinning, so no energy is lost due to friction between the axle and the string. When the yo-yo starts to spin more slowly as it runs out of energy, springs press on the clutch so the discs and the axle spin together. Now, the spinning axle winds up the string and the yo-yo climbs up. You have to wonder what Flores would have thought of this innovation.

Inside the Yo-Yo

We bought a Yomega Power Brain Yo-Yo to take apart. It has a clutch system. The two discs unscrew. All the interesting stuff is in one side, and you can see it through the translucent cover. But we took it apart anyway. We cut a small hole into which to slide a screwdriver to pry up the cover. It took a few minutes, but the cover eventually popped off. Inside are two springs that clamp plastic arms onto the plastic axle when the yo-yo slows down. Metal ball weights fling outward at high spin rates, letting the yo-yo sleep until it slows down. It takes an impressive force against the springs to push open the clutch.

Build Your Own

It's easy to make at least a simple yo-yo. We've used two 2-inch diameter wood wheels connected by a short piece of ¼-inch dowel. Heavier wheels would be much better. You can purchase bags of yo-yo components ready for assembly at some craft and hobby stores.

ball weights

springs clamp onto axle

axle

string groove

Try different types of string to see which gives optimal friction. Tie a loop in the end of the string and slip this over one wheel and onto the axle. The knot must be small; otherwise, it catches on the string.

Resources

Woodworks Ltd. (www.craftparts.com) sells bags of yo-yo components, including dowels, heavy wheels, and string.

BIBLIOGRAPHY

Anderson, David F., and Scott Eberhardt. *Understanding Flight.* New York: McGraw-Hill, 2001.

Baldwin, Neil. *Edison: Inventing the Century.* New York: Hyperion Press, 1995.

Balmer, Al, and Mike Harnisch. *Mouse Trap Cars: A Teacher's Guide.* Round Rock, TX: Doc Fizzix Publishing Company, 1998.

Ehrlich, Robert. *Turning the World Inside Out and 174 Other Simple Physics Demonstrations.* Princeton, NJ: Princeton University Press, 1990.

Francis, Neil. *Super Flyers.* Reading, MA: Addison-Wesley, 1988.

Gilbert, A. C., with Marshall McClintock. *The Man Who Lives in Paradise: The Autobiography of A. C. Gilbert.* New York: Rinehart and Company, 1954.

Hoffman, David. *Kid Stuff: Great Toys from Our Childhood.* San Francisco, CA: Chronicle Books, 1996.

Hollander, Ron. *All Aboard: The Story of Joshua Lionel Cowen and His Lionel Train Company.* Revised and updated ed. New York: Workman Publishing Company, 2000.

Lorenz, Ralph D. *Spinning Flight: Dynamics of Frisbees, Boomerangs, Samaras, and Skipping Stones.* New York: Springer, 2006.

Panati, Charles. *Extraordinary Origins of Everyday Things.* New York: HarperCollins, 1989.

Ruhe, Benjamin, and Eric Darnell. *Boomerang: How to Throw, Catch, and Make It.* New York: Workman Publishing Company, 1985.

Sobey, Ed. *Fantastic Flying Fun with Science.* New York: McGraw-Hill, 2000.

Sobey, Ed. *A Field Guide to Household Technology.* Chicago: Chicago Review Press, 2006.

Sobey, Ed. *A Field Guide to Office Technology.* Chicago: Chicago Review Press, 2007.

Sobey, Ed. *A Field Guide to Roadside Technology.* Chicago: Chicago Review Press, 2006.

Sobey, Ed. *How to Build Your Own Prize-Winning Robot.* Berkeley Heights, NJ: Enslow Publishers, 2002.

Sobey, Ed. *Inventing Toys: Kids Having Fun Learning Science.* Tucson, AZ: Zephyr Press, 2002.

Sobey, Ed. *Loco-Motion: Physics Models for the Classroom.* Chicago: Zephyr Press, 2005.

Sobey, Ed. *Rocket-Powered Science.* Tucson, AZ: Good Year Books, 2006.

Sobey, Ed. *Wacky Water Fun with Science*. New York: McGraw-Hill, 2000.

Sobey, Ed. *Young Inventors at Work: Learning Science by Doing Science*. Glenview, IL: Good Year Books, 1999.

Walker, Jearl. *The Flying Circus of Physics*. New York: John Wiley, 1977.

Walsh, Tim. *Timeless Toys*. Kansas City, MO: Andrews McMeel Publishing, 2005.

Zubrowski, Bernie. *Raceways: Having Fun with Balls and Tracks*. New York: William Morrow and Company, 1985.